W9-BAX-063

JAZZ
IN ITS TIME

Martin Williams

New York Oxford
OXFORD UNIVERSITY PRESS
1989

Oxford University Press

Oxford New York Toronto
Delhi Bombay Calcutta Madras Karachi
Petaling Jaya Singapore Hong Kong Tokyo
Nairobi Dar es Salaam Cape Town
Melbourne Auckland

and associated companies in
Berlin Ibadan

Published by Oxford University Press, Inc.,
200 Madison Avenue, New York, New York 10016

Library of Congress Cataloging-in-Publication Data
Williams, Martin T.
Jazz in its time / Martin Williams.
p. cm.
Includes index.
ISBN 0-19-505459-8
1. Jazz music. I. Title.
ML3507.W535 1989
785.42—dc 19 88-31954 CIP MN

The following page is regarded as an extension
of the copyright page.

2 4 6 8 9 7 5 3 1

Printed in the United States of America
on acid-free paper

Acknowledgment is hereby made of permission to reprint in somewhat different form essays first published by the following:

down beat: Bechet the Prophet; Two About Pee Wee; One About Lester; Three About Coltrane; Jazz at the Movies; Miles Davis Live; Errors of Interest; Roaring; Mr. Wilson; Navarro out of the Air; Condition Red; Record Date: Art Farmer and Jim Hall; Blues Night; Bash It; Criticism. By permission of *down beat* magazine.

International Musician: Lee Konitz: A Career Renewed; Lionel Hampton: Major Contributions; Bud Freeman: The Needed Individual; Thad Jones: A Musical Family; Bobby Hackett: Everything with Feeling; Harry Carney: Forty-one Years at Home; Bash It.

Jazz Times: How Long Has This Been Going On?; Monk Goes to College.

Metronome: A Rock Cast in the Sea.

Musical America (High Fidelity): Biographies, Autobiographies, Profiles, and Oral History.

Saturday Review: Bechet the Prophet © 1967; Two About Pee Wee © 1965; Three About Coltrane © 1965, 1966; Jazz at the Movies © 1967. All copyright by *Saturday Review* magazine.

Sonneck Society Bulletin: Grove American I: Not Just Missing Persons.

In memory of J.E.C. and
again for Mark, Rick, and Doug

Introduction

For a writer to re-read his early reviews and occasional columns can be a humbling experience—and it should be. Early reviews that don't seem foolish or wrong-headed may seem poorly written. And those that don't seem so wrong-headed, and seem well enough written, may not have as much substance as one thought at the time. I hope the reviews and columns I have included in this book will seem neither misguided nor poorly composed to the reader.

Jazz in Its Time is both like and unlike my previous book, *Jazz Heritage*. It contains more short reviews and more short columns than that volume. And I have included a somewhat lengthy effort to evaluate developments during the music's past two decades. The book also offers some brief profile-appreciations of musicians.

Some of the reviews and columns grouped themselves rather naturally by subject. There are, for example, two views of Sidney Bechet and two on jazz movie scores, written several years apart. There are three comments on John Coltrane written across a seven-year period. And the piece titled "Bash It" combines two rather different approaches to Jaki Byard and his music.

There are some LP record annotations included, and they group themselves in seven accounts of important trumpeters, and three comments on Ornette Coleman's early recordings. Also included are narrative pieces recounting recording dates, an evening in a night club (a rather unusual one), and a rehearsal (again, an unusual one). (Would these be examples of the "new journalism"? I think not, but then I think that term itself may be misguided.) Finally, there are some reflections on the nature of criticism and the state of jazz scholarship.

Aside from making some verbal changes, some clarifications, and perhaps a *bit* of discreet second-guessing, I have left things pretty much as they were when these pieces first appeared, with no effort to update them, and only a few added postscripts, all of that—as indicated by the book's title—to reflect the music and the writing about it in its time.

As the acknowledgments and credits reveal, these writings originally appeared chiefly in *down beat,* the *Saturday Review,* the *International Musician, Evergreen Review, Metronome, Musical America, Jazz Times, Jazz Educator's Journal, American Music, The Sonneck Society Bulletin, FM Guide,* and on the back liners to LP jackets from Atlantic, Milestone, Prestige, and the Smithsonian Collection. I am grateful to all concerned for their first appearances and for permission to include them here.

Alexandria, Virginia
1988

Martin Williams

Contents

I

LISTENING

Bechet the Prophet

In France in the 1950s, New Orleans clarinetist and soprano saxophonist Sidney Bechet found himself a celebrity: his records were collected by ardent teen-age fans; his company was sought by socially ambitious matrons; and his night club and music hall appearances were as avidly attended as those of Chevalier or Piaf. In our country probably only Louis Armstrong among jazzmen has received comparable adulation. But Bechet did not sing, did not clown; he played his instruments, accompanied by French New Orleans "revivalist" ensembles of highly uneven quality, and he announced his numbers in the softly accented Creole French of his New Orleans upbringing. Armstrong, the glorious instrumentalist of the New Orleans school, was something of a prophet without honor at home to larger audiences until he became the grand old man. Bechet remained without honor at home to even a sizable audience. Yet among instrumentalists, he was the second greatest glory of New Orleans music and her greatest reed man.

Bechet was recognized as a major jazz musician outside his own city by the teens of this century. He was the subject of the first serious appreciation ever written about a jazzman, conductor Ernest Ansermet's tribute of 1919. Ansermet saw in Bechet's work the beginnings of a new style and he conjectured that tomorrow perhaps the whole world would be following his path. We also learn from Stravinsky's letters that he wrote three early clarinet pieces under the spell of Bechet. During subsequent developments in jazz, Bechet was admired by musicians of all styles and schools; he was one of Charlie Parker's favorite players, and one of his last recordings—an excellent one—was done with the respected modernist French pianist, Martial Solal.

In Antibes there is a square named after Bechet with a bust of the musician in its center. It would probably be foolhardy to expect such a thing in this country, but perhaps one can hope at least for an adequate posthumous tribute to his music in the form of LP reissues, for, although he remained a major player to the end, he did some of his best playing here and made most of his best recordings here, and for a jazzman, records are the enduring statements of his work.

Bechet made some mature recorded statements between 1938 and 1941, many of them for Victor. Those recordings show him to be everything that a great jazz musician of his generation might be, and more. Like the best clarinetists of his home city, he could improvise a superb ensemble part (*I Ain't Gonna Give Nobody None of This Jelly Roll*). And because his approach on soprano sax combined elements of both the trumpet and the clarinet traditions, he could provide a unique lead part in ensemble as well (*Shake It and Break It*). He could discreetly interpret a lyric theme (*Indian Summer*) or more boldly paraphrase a line (*Mood Indigo*). On occasion, he might even invent melody only within the harmonic guidelines of a piece (*Sleepy Time Down South*), a rare practice for a musician of his era. Above all, he was a great blues player (*Texas Moaner, Nobody Knows the Way I Feel This Mornin'*), and that means also that he could use the jazzman's growls, smears, bent notes, shouted notes, and whispered notes with innate aesthetic discretion in any context.

When he growled or shouted or laughed, Bechet sang. And if the paradox of simultaneous shout and song were not enough, he was at once harsh and tender, introspective and declamatory, personal and communal, melancholy and joyous, and never egocentric. "The music," he said, "that's a thing you gotta trust. You gotta mean it." The words come from his book *Treat It Gentle,* which, for at least half its length, is one of the most remarkable autobiographies ever written by an American artist.

Bechet's Victor recordings were made with pick-up groups; their size and instrumentation varies from trios through septets, and the approaches and styles of the players roam widely for the time—his drummers, for example, included traditionalist Baby Dodds and then-modernist Kenny Clarke, his trumpeters Sidney de Paris and Charlie Shavers. For a listener, one of the revealing pleasures in these recordings is Bechet's adapting not only to the range of the selections of each date but also to the sizes and shapes of the ensembles and the personalities of his fellow improvisors—and sometimes on each of his two horns. Bechet accomplishes things which only a player of his technical and emotional largess could accomplish.

Thus Bechet's Victors present the strongest argument for an orderly LP presentation ensemble by ensemble, record date by record date. Alas, the two LPs in RCA Victor's Vintage series, *"The Blue Bechet"* and the earlier *"Bechet of New Orleans,"* do not follow this obvious and easy procedure. Rather, they follow some strange principle of separating blues from stomps, a principle which does not work in practice—more than half the pieces on the *"The Blue Bechet"* are *not* blues, but one of the best performances on *"Bechet of New Orleans"* is. (The albums provide further disappointments: neither contains Bechet's *What Is This Thing Called Love,* which has never been issued in this country and which is quite possibly his greatest reading of a standard ballad, and the albums mistakenly claim to present a previously unissued "take" of *Sidney's Blues.*) These reissues do give us a great musician at the peak of his greatness, and all of the above examples of his range come from just two Vintage LPs.

I would like to single out another performance from the second Vintage set, *Blues in Thirds*. It was made during a Chicago visit in 1940 by a trio consisting of Bechet, pianist Earl Hines, and drummer Baby Dodds. As I have remarked before, Hines is no bluesman, but he showed in this piece (his own) a typically lyric modification of the blues idiom. And here that lyric mood is reinterpreted by a master bluesman, Bechet. Hines, in his two opening choruses, states his theme superbly and then provides a variation. Bechet enters, first with a return to Hines's melody, then with a powerful, contrasting Bechet blues invention. He continues with fresh improvisation in the final chorus, while Hines counterpoints the opening theme.

For Bechet, the performance exhibits nearly all the powers I have attributed to him—powers of paraphrase, powers of invention, powers of adaptation to a particular musical climate—plus an outstanding sense of overall structure. *"Blues in Thirds"* could easily be Bechet's masterpiece on clarinet, I think; indeed for both men, and possibly all three, it is a superb performance. This is its first American reissue and hence its first issue in twenty-six years. That is at least some kind of tribute. *(1967)*

It happened this way: Sidney Bechet, who has been a major figure for over thirty-five years and has the respect of musicians from every style, heard pianist Martial Solal, who has been called the most advanced jazz musician in France, and wanted to record with him. Two record dates resulted, one with Kenny Clarke on drums.

On the face of it, one would expect clashes—rhythmically, harmonically, even emotionally. Bechet's rhythmic conception has grown less like, say, King Oliver's and more like middle swing through the years. Solal is, well, Solal is rather like a cross between Hank Jones with touches of Oscar Peterson and Monk (does that make sense?). By the time the two of them are trading fours on *These Foolish Things* (the first track made), it is evident that the whole thing is going to work splendidly.

There is a mutual agreement on the things that matter and a mutual joy in creating music. There is also something which the comparative brevity of each track dramatizes: the terse condensation and completeness of statement that each of these men is able to make—a lesson to many extended blowers (and a&r men). Bechet is not always at his very best but he is always authoritatively alive, and both *It Don't Mean a Thing* (on which Clarke is splendid and Solal witty) and *The Man I Love* may well stand comparison with his finest performances. Solal's is an exciting talent and one firmly in hand. It is also quite wonderful to hear the way each of these men will bow to the other by echoing a phrase or device and then proceed to go his own way.

This is neither a curiosity nor a stunt but a respectful meeting of individuals. It is also a lesson for anyone who has prejudices about jazz styles. If there had been a blues included, it would have been (as always) a fine vehicle for Bechet and an interesting test for Solal. (*1958*)

Two About Pee Wee

A new release on Counterpoint, *"A Portrait of Pee Wee,"* shows us (again) that Pee Wee Russell takes chances, dares, explores. He is never intimidated by a melody or a chord structure. Even when he uses devices that are common in his work, say, on *I Used to Love You* here, he is not playing it safe, coasting, or being trite.

That almost dramatic adventurousness is disciplined by a genuine, even classic and balanced lyric-melodic gift, of which his solo here on *Out of Nowhere* is a very good example. And years ago that same spirit intuitively led him into harmonic explorations far ahead of their time. Even in relative failure, he can be interesting.

He is, of course, original and individual. One might almost say that he adapts music and his instrument to himself, not himself to it. But any discussion of his instrumental technique is obviously beside the point; he is thoroughly musical in the *real* meaning of that phrase.

He is not presented here as a "Dixieland" musician, which is good because he was never really a good polyphonic player; he is a soloist. Nat Pierce's arrangements, although a couple scored in the manner one might use for a larger group, acknowledge this and acknowledge the fact that his approach is primarily melodic and not rhythmic.

The others present include trumpeter Ruby Braff, tenor saxophonist Bud Freeman, and trombonist Vic Dickenson, and they are not resting on past achievements either. But for this time let the report be on Pee Wee Russell. *(1958)*

If you had asked an avid and "inside" follower of jazz about clarinetists in the late thirties or early forties, he might well have answered—with due respect to the public successes of the day and particularly Benny Goodman—that there was a fellow named Charles Ellsworth ("Pee Wee") Russell, who spent most of his time with the displaced "Chicagoans," working in and out of Nick's and Eddie Condon's in Greenwich Village.

Russell's work probably came as something of a shock for those who heard him for the first time, even if they knew his name and reputation. Self-evidently, a clarinetist like Goodman could, with only slight adjustments in his reed technique, slip from an exposition on *Stompin' at the Savoy* to the Mozart Clarinet Quintet. But Russell, with a series of growls, phlegmy undulations, and occasional off-pitch peals, had converted this classically complex instrument into a personally expressive extension of Pee Wee Russell. A raspingly eccentric Dixielander, some thought. Far from it, for every unorthodox sound is a part of Russell's technique and the basis of that technique is always firmly musical.

Russell was virtually lost, like so many of his generation, during that lamentable decade when jazz was either old New Orleans or new be bop, with little or nothing in between, but he has lately experienced a remarkable rediscovery. It may surprise those now middle-aged jazz fans who knew him when, and who no longer read the fan and trade magazines, but Pee Wee Russell now places high in the annual popularity polls. He deserves the rediscovery, as his recent work testifies. He also deserved his earlier "inside" reputation, as a few of recent reissues similarly demonstrate.

The first of these is *"Pee Wee Russell—A Legend."* The LP is built around a series of trios and quartets that Russell recorded for Commodore, an "inside" jazz label of the period. On the face of it, trios and quartets are apt to present Pee Wee Russell more effectively than the frayed, Condon-engineered Dixieland groups with which he often recorded for that label, for Russell is pre-eminently a soloist, not a heterophonic ensemble improviser. But the trios available are not ideally handled on the recent Mainstream reissues of his work on Commodore, for two of them (*Jig Walk* and *About Face*) are not included, whereas *Deuces Wild,* in which Russell provides moody relief on a vehicle for Zutty Singleton's tom-toms, is.

Not that the selections included all feature ideal performances. On *Back in Your Own Backyard* and *Rose of Washington Square,* an apparent unfamiliarity with the material has both Russell and pianist Jess Stacy falling back on their own pet phrases. However, on *Keepin' Out of Mischief Now,* each man comes up with striking and often amusing ideas in paraphrase of Fats Waller's melody. Similarly, in his solo with a larger group on *A Good Man Is Hard to Find,* Russell is so full of ideas and so effective in his handling of stop-time breaks that the original melody virtually (and justifiably) disappears.

A performance like that one leads us to a second test for a major jazzman: Can he extemporize a good melody without relying on paraphrase? There are two blues included in the LP on which Russell does that. On *D.A. Blues* he offers some touching lyric musings, perhaps under the influence of Stacy's lovely piano choruses and his gentle touch. On *The Last Time I Saw Chicago,* with the sharper and more percussive touch of pianist Joe Sullivan, Russell muses *sotto voce* over keyboard tremolos and then expansively proclaims his conclusions.

The LP is filled out with three other selections by large ensembles in which Russell happened to be present, and, aside from a complaint about Mainstream's usual overall skimpiness of playing time, there is again a question of selection. Two of Russell's grandest recorded statements, both made for Commodore, do not appear here. The first of these is his half-chorus on *Embraceable You,* a truly exceptional invention that barely glances at Gershwin's original, tellingly placed among the statements of Bobby Hackett, Bud Freeman, and Jack Teagarden, which by contrast variously embellish and ornament the original. This performance found its way into a catch-all Mainstream anthology called *"Era of the Swing Trumpet,"* as did Russell's equally inventive solo on *Lady Be Good,* which is dropped into *"Eddie Condon: A Legend."* (*1965*)

(Subsequent reissues of the Commodore material included many of the selections discussed above and appeared on a restored Commodore label in the U.S. and a "Commodore Classics" series of imports.)

One About Lester

If masterpieces exceed excellence, then there is not a high enough rating for the Lester Young on the Commodore recordings by "the Kansas City Five."

A 1938 session (tracks 1–6, 11, 12 of the LP) produced the masterpieces: his delicate contrapuntal clarinet improvising behind Buck Clayton and his tenor solo on *Way Down Yonder in New Orleans,* his delicately original clarinet invention on *I Want a Little Girl,* and his lovely variant of a traditional blues line on *Pagin' the Devil.* These are masterpieces.

Excellence comes with his clarinet solo in the faster *Countless Blues,* his interplay and tenor invention on *Them There Eyes.*

Surely no horn man between Louis Armstrong at his 1933 peak and Charlie Parker beginning in 1945 produced more original or more beautiful work than the best of Lester Young, and these solos are among the best. And as surely as the best jazz does survive its time and period, these do. Astonishing: he was a great tenor player, yet for the way he handled the clarinet's sound alone, he would be an exceptional clarinetist, and of the handful of clarinet records he made almost all are gorgeous creations.

There are other things to notice. Buck Clayton is fine, for instance; Eddie Durham, the leader and a good arranger, has such a very different and older sense of time on guitar from the other men. And notice the group: horns and rhythm with no piano—the idea has been tried since and less casually than this but not really more successfully.

But then, masterpieces never are surpassed. The only thing to do with masterpieces is hear them and treasure them and—if they are jazz—be thankful they got recorded. (*1962*)

(*The session discussed here has been reissued several times since, most recently in the U.S. on a revived Commodore label.*)

Three About Coltrane

"*Mainstream 1958*" by Wilbur Harden, John Coltrane, and Tommy Flanagan is a kind of ersatz Miles Davis date. And that quality goes beyond the presence of certain musicians (Doug Watkins on bass is also here) and the fact that Harden is following Davis's style quite closely: *Snuffy* (Harden's tune) is a blood brother to such things as *Little Willie Leaps* and *Half Nelson* from Davis's first date as leader. But Harden deserves credit for largely getting the point of what Davis does and doing it better than some others—I am thinking particularly of some of the West Coast men who play Davis's ideas but who patch them together so incongruously (even putting some climactic ones in their first eight bars) that they sound almost like a parody of the original.

In trying to decide why Coltrane's runs of short notes don't sound like double-timing, I got the idea that he may be working on a new subdivision of the rhythmic concept of jazz, one which further divides the eighth-note unit of be-bop into a sixteenth-note rhythmic conception.* And

* I later decided I was wrong here. See below, p. 54.

the fact that in him this attempt is complemented by a deeper harmonic approach suggests a sound evolutionary balance of the kind that bop had. If this is so, with such a task undertaken, it is little wonder this his is still largely a vertical style and that he has not yet arrived at any real discipline of form. Here he seems to be using that rhythmic conception more conservatively (his solo on *West 42nd Street* has a commendable rhythmic development as a result) which may be the way he will use it eventually, both for his own development as a soloist and for use in a specifically melodic playing. But such conjectures perhaps only indicate how exciting the prospects are, and patience for all may be the best thing to suggest. When the plant is growing, it doesn't do to keep pulling it up to look at the roots.

On Tommy Flanagan, I can repeat what I've said before about what a real pleasure his straightforward and inventive way of playing is after one has heard the way some others hoke up a style of comparable conception with obvious cocktail trickery.

The notes spoil for a fight about whether these men make "mainstream" jazz as of 1958. No arguments from here: streams flow. They aren't ponds or puddles. (*1958*)

Charles Lloyd is not the first player to emulate John Coltrane of course—hundreds of young men have been doing that for years now. But in a sense he may be the last, for the John Coltrane heard on the LP called "*Crescent*" is a rather different musician. I think it is his best album in several years, and it is at least an exceptional set and an excellent presentation of his work.

It is hard to believe that the poised melodist of this LP is the same Coltrane who once seemed to gush out every possible note, to careen through every scale on every complex chord—and go beyond even that profusion by groping for impossible notes and sounds on a tenor saxophone that seemed about to burst under the strain. Or who, to bring order to this swirl and fury of notes, subsequently cut back the harmonic structures of his pieces so drastically that some performances sounded like vamps-till-ready, or perhaps twenty-five minutes of furious E-flat sevenths.

I doubt that the earlier Coltrane would have played the lovely incantive melody that introduces *The Drum Thing* on this new album without embellishing it out of shape. The performance also includes an exceptionally self-contained and satisfying drum solo by Elvin Jones, which at the same time does not sacrifice his formidable technique. On *Lonnie's Lament* the saxophonist similarly sets up solo space for his pianist McCoy Tyner, who here has his harmonic sense under control, and for his bassist, Jimmy Garrison, who offers a gratifyingly personal statement.

On the three numbers that feature the leader more predominantly, Coltrane is like a man returned from a dangerous but mandatory journey, now ready to share the experience. Strange beauties and perils are vividly related in the improvised section of the title piece, *Crescent;* even the once "impossible" notes and sounds have become natural and firmly established musical techniques for him. Musings and evaluations of the journey take place on *The Wise One*. And *Bessie's Blues* is like an up-tempo, communal celebration of the new experiences. (*1965*)

Tenor and soprano saxophonist John Coltrane is a successful musician, but at the moment his position does not seem enviable. His initial popularity depended on a hit version of *My Favorite Things* and he is required not only to perform that piece often but to follow it with readings of similar material. He feels responsible to assist and encourage a segment of younger "advanced" musicians who consider themselves his followers. At the same time, he is under attack from reviewers who do not like Coltrane's work or its implications and often end up blaming him for a kind of wilful musical obscurity. More recently Coltrane's music has become the center of a nasty journalistic in-fight over its "content" and "meaning" and its relationship to Black Nationalism—a controversy in which Coltrane's supposed "meaning" turns out to be some opportunistic Marxist clichés delivered with a somewhat Mao-ist accent.

It has been a little over a year since I wrote of Coltrane and a backlog of several releases has accumulated by now. *"A Love Supreme"* comprises a single, four-part piece by the Coltrane Quartet. It is offered in a rather austere black-and-white jacket with liner notes by Coltrane himself. This unusual LP was barely released before it was selling very well, and it has already won a couple of popularity polls. The notes begin "All praise to God" and programmatically the piece concerns a religious experience, a period of irresolution, and a return to faith. Each section opens with a theme, followed by "free form," modal improvisation by Coltrane and the members of his group. Considering the subject matter and the strong emotion usually generated by this ensemble the general tone here is relatively calm.

By contrast the frugally titled *"John Coltrane Quartet Plays"* is offered in a full-color jacket and would seem to have fairly conservative, even commercial, intentions, including versions of *Chim Chim Cheree* and *Nature Boy*. The explorations, however, are sometimes quite strong technically and emotionally.

"Ascension" is the most daring recording Coltrane has ever made. It is a continuous thirty-eight-minute performance in which Coltrane's quartet is augmented by two trumpeters,

two tenor saxists, two altos, and an extra bassist. It utilizes a single, slight thematic idea, several loose, turbulent improvised ensembles, and solos by most of the players. It blares, rages, shouts, screams, and shrieks. It alo soars and it sings. It is at once a truly contemporary performance and a kind of communal rite.

There are surely many things to admire in these records. There are some of Coltrane's ingenious spidery lines on *Chim Chim Cheree*. There is drummer Elvin Jones's inspired playing on the same piece. And although I have felt that pianist McCoy Tyner's harmonic sense was overly lush for this music, his solo on *Brazilia* on the *"Quartet Plays"* set is hard and gem-like.

However, some of this music seems to me repetitious, and there are moments when Coltrane's wildly authentic passion seems not so much a part of the music as a part of the musician—the reaction of a player who is improvising with a minimum of built-in protection but who sometimes cries out, frustrated, against the very challenges he has set for himself. Often, one's final impression is of musical statements that are highly charged and have brilliant moments but that are somewhat static and unresolved—statements sometimes contained only by a fantastic and original saxophone technique on one hand, or by a state of emotional exhaustion on the other. (It is surely indicative that many of these performances are faded out mechanically rather than ended musically.)

Certainly Coltrane's music is related to the mood of American Negroes, and particularly the awakenings and frustrations of young American Negroes—as if anyone ever doubted that it was, or as if it could be meaningful to so many people if it were not. And in the sense that it is, I think a performance like *Ascension* might be heard, felt, and reflected upon by every politician, police official, psychologist, social worker, editorial writer—perhaps every American—whether it succeeds aesthetically or not. If a listener is at first confused or repelled by it, perhaps he should hear it again. To put it another way, if there were a documentary film on the Watts riots, I think John Coltrane would be the ideal man to score it—and I intend the remark to characterize and compliment his work, not to criticize it.

The challenge of America's racial problem is, as the young James Baldwin saw so clearly, a fundamental challenge to Western civilization and all its traditions. And Coltrane's jazz, like much good jazz, looks deep into the inner-being of all men. Some musical statements such as his, even if they fail aesthetically, are certainly not to be dismissed.

If *Ascension* leaves one with a feeling of despair, he might turn to Ornette Coleman's *"Free Jazz"* (Atlantic 1364), a performance to which *Ascension* is directly indebted, but for me a work of beauty and affirmation and hope. For me *Free Jazz,* in Joyce's phrase, better sees the darkness shining in the light. Still, it is a sight for which Coltrane can prepare us, and the preparation can be invaluable. The aforementioned *Brazilia* stays with one and echoes through one's being long after its notes are spent. (*1966*)

Jazz at the Movies

· I ·

"One Never Knows," by the Modern Jazz Quartet, offers music composed by John Lewis and played on the soundtrack of a confused and trashy French sex-pot movie (without Brigitte Bardot), *Sait-on Jamais (One Never Knows,* but

called *No Sun in Venice* by the American distributors) and even with its faults it represents an achievement in several respects.

The role of sound-track music is, of course, entirely functional. Its basic purpose is to complement the film, comment on its action and mood, and (at base) keep the audience from becoming distracted. The moment the usual film score begins to draw attention to itself, it fails. Despite the fact that in the film one hears only what is in effect snippets of the score, I think most of it can stand (with Virgil Thomson's *The Plow That Broke the Plains,* Prokofiev's *Alexander Nevsky,* and, perhaps, Aaron Copland's *Of Mice and Men*) as exceptional, as a film score that ingeniously manages to be both effective in context and strong enough to stand on its own.

Second, it is now evident that the group has so thoroughly assimilated and transformed the eighteenth-century fugal form, that we can no longer speak of a pastiche, a novelty, or a toy, but, for the Quartet, of a jazz fugue. And perhaps that means that the kind of specifically jazz polyphony heard here on *The Rose Truc* (a blues, despite what the liner says), something John Lewis and Milt Jackson do so excitingly, can now also continue to develop.

The two fugues, *The Golden Striker* and the triple fugue *Three Windows,* dramatize the fact that the slight rhythmic disunity heard on the quartet's previous release is not a fact and may have been a flaw in audio engineering. Connie Kay's work is almost unbelievably integrated. He is magnificent on *The Golden Striker* and manages to overcome the role handed to him on *Three Windows,* which might have become monotonous.

One Never Knows is a lovely melody and shows again how much Lewis can make out of the simplest materials, but the performance explores it little, using it almost as a vehicle for dynamics and tone colors. The same sort of thing happens on *Cortège,* which is rescued a bit too late from a lushness and returned to it too soon. At this point in its career, the Quartet still seems to find it necessary to treat such exhibitions as if they were ends in themselves—which they are not. *Cortège* (another blues) begs comparison to the Quartet's (and Lewis's) masterpiece, *Django,* and *Django* succeeds by an opening and closing condensation of theme-statement and melodic exploration where *Cortège* fails. *Venice* is another simple melody which a solo by Jackson rescues from a certain cocktail-ish impressionism; Lewis's solo therein is not up to the level of passion understated which he achieved on *The Bad and the Beautiful* in his solo album, *"The John Lewis Piano."*

Finally, the recording preserves a better performance of the score than any of several recent "in person" ones that I have heard. (*1958*)

· II ·

What is a good film score? Is it music which is good enough to stand on its own, out of its original context? Or is it, on the other hand, music which is effective in context but which might not stand up in a concert hall?

On film, effective drama is a complex of several things. Movies with sound and dialogue are still primarily visual experiences like their silent forebears. But they begin as written narratives, which portray character, situation, and plot through dialogue. When they are executed, they also include photography, a carefully edited montage of "shots," and some "background" music as a part of their final effect. At the crudest level, one might say that the music is there simply to keep the audience from becoming distracted. And at another level, it is there to underline and perhaps complement mood, situation, and character.

The film composer walks a narrow line; he has to be good

enough not to be noticed. If he does not do his job well, he will be noticed, either because he does not contribute to dramatic effect well enough or because, one might say, he contributes too much—he distracts one from the drama and draws too much attention to himself. We would not underline a dramatic film with a Beethoven symphony because, no matter how good the film, the audience might end up listening to Beethoven. In short, good film music is a purely functional aspect of one kind of drama.

One example of ideal movie music would be Max Steiner's score for *King Kong*. It sounds perfectly marvelous, is almost totally effective in context, and yet one probably could not bear it for more than five minutes if he were asked to sit attentively during a performance in a concert hall. There are several film scores, however—one thinks particularly of a couple of Virgil Thomson's and Aaron Copland's, which manage to sound right in context and yet do stand up on their own.

About ten years ago, the movies began an earnest use of jazz as background music. Naturally, the industry discovered some fairly derivative hacks of its own to grind the stuff out, but it also invited some more-than-capable jazz musicians, and even some first-rate ones, into the fold.

It may seem odd that jazz, a form of music which has been so long determined to free itself from a purely functional role as a background in the barroom and dance hall, should have embraced so enthusiastically still another functional role as background at the movie house. But jazz, as usual, proved to have an unpredictable vitality. John Lewis wrote some very good pieces for the Modern Jazz Quartet to play in the background during a French movie called, in this country, *No Sun in Venice*. And Lewis wrote a first-rate score for a small orchestra for *Odds Against Tomorrow*. That latter score, paradoxically, was even more functionally conceived in terms of the action of the film, but was somewhat better heard on its own.

Perhaps the most interesting approach was used by Miles Davis for an entirely forgettable French new-wave film called, over here, *Frantic!* Davis and his sidemen watched a screening of the movie and made some notes and sketches. Then they proceeded to improvise their score in direct response to a second screening of the movie in the recording studio. (*1967*)

Miles Davis Live

Miles Davis's *"In Person, Friday and Saturday Nights"* was recorded at the Black Hawk in San Francisco. I am reviewing it as a two-record set, but each LP is available singly, as *Friday Night,* consisting of the first six titles above, and *Saturday Night,* the last six.

The recording is generally very good, except that, to my ears, Davis's intimately close-miked sound is on the verge of distortion (without quite making it) on *Fran-Dance* and *If I Were a Bell.* One other flaw is also apparently technical, a decided drop in tempo between Davis's opening solo in *All of Me* and Wynton Kelly's solo—a tape splice it would seem.

When Davis is good here, he is good indeed, and the only places where I think he is not really good are on *Fran-Dance,* on which his lower register clouds up and he gets hung on a single lick, a cliché in fact, which he somehow never executes cleanly. And more or less the same idea hangs him on *Bye, Bye Blackbird,* though only briefly. Those are the only places where fluffs bothered me. Also, I have never thought much of *Fran-Dance* (*Put Your Little Foot Right In*) as material, and this seems to be a rather diffuse performance of it.

This visit to *All of You* seems diffuse; Davis's first solo brings up some good ideas but never really finds its direction. His return at the end rebuilds things excellently, however.

On *All of You,* pianist Wynton Kelly is good. He tries for less than Davis, but his playing has direct organization. I think that, next to Davis, Kelly is the soloist here. I particularly admire the way he has fallen into a somewhat preassigned, Ahmad Jamal-like role in this group. Even when his ideas are not exceptional, they are usually good, and he always delivers them with personal force and conviction. He has an interesting solo on *Oleo,* with good variety of phrasing, fleetly delivered, and his solo on *If I Were a Bell* is marred only by some rather predictable phrase lengths that he employs toward the end. However, I confess I can make little out of *Love, I've Found You,* which features Kelly alone. Granted that it is very well done, but such an out-of-tempo version of a pop tune, confined to a one-chorus statement with a few embellishments and some altered chords, seems to me to belong under the conversation in a cocktail lounge, no matter *how* well done.

Tenor saxophonist Hank Mobley is capable in this company. When Kelly lays out behind him on the very fast *Walkin',* for example, his time seems to falter. My first impression of *Walkin'* was that it was an impatient version, played fast because of the boredom of having to answer requests for a hit record night after night. I was wrong; it is not. Davis is on top of the tempo, and he plays very good blues of his own special kind, with a wonderful climax to his solo. And if there is one lesson that his phrasing could teach, it is that one needn't clutter up a solo with notes at *any* tempo, that if one concentrates on melody and continuity, his lines can be simple, his pauses eloquent.

I don't want to belabor the point, but Mobley runs some fairly standard phrases on *Bye, Bye, Blackbird,* including some out of Sonny Rollins, and on *So What,* which is of course built on two modes rather than chords, one is soon conscious of the underlying mechanical framework during Mobley's episode.

On *Neo,* Mobley is emotionally compelling from the first, but before he is through, he has played almost all those flamencan phrases that usually show up on the sound tracks of pictures about bullfighters.

On that same *Neo,* which is more or less out of Davis's LP *"Sketches of Spain,"* Davis is eloquent, almost as movingly eloquent as he was on *Saeta*—and that means that he is almost as eloquent as any jazz musician is likely to get. The materials of *Neo* are very simple, and the temptations to run Latin clichés or fall into a monotony of sound, melody, or emotion are enormous. Davis gives in to them not at all.

I was fascinated by *Bye, Bye, Blackbird.* Davis's theme statement has become a tantalizing, suggestive sketch of the original. There is an effective little modulation now, and the trumpeter's variations are better here than on any version of this piece I have ever heard him play. He also has new ideas on *If I Were a Bell.* I did feel, however, that *Oleo* has become a bit too fast for its own good.

No Blues has a medium funky line, taken at almost perfect tempo for the melody and the kind of variations that Davis comes up with. His emotional rage in this performance is something to hear, all the way from his own version of simple earthiness through the kind of blues lyricism at which Miles Davis is unique and including some humorous, "corny," on-the-beat licks en route.

Best of all, for me, is this fast version of *Well, You Needn't.* I have not heard such sprightly, nearly breathless, and original rhythmic interest from Davis since that superb solo on *Boplicity.* I play his section of *Well, You Needn't* in delight and almost in disbelief. I shall remember his ideas there and on *Bye, Bye Blackbird,* and his eloquent speech on *Neo,* for a long time.

I hear a lot of people in this man's work. Foremost I hear *him,* unmistakably. I also hear Lester Young, I hear Freddy Webster, and I hear ideas of Dizzy Gillespie and Charlie Parker transmuted and put to such very different use that they are almost unrecognizable.

I hear a personal use of sound that also sometimes suggests that Davis is trying to reinterpret the whole range of sound of the Duke Ellington trumpets of 1939—Cootie Williams's plunger and Rex Stewart's half-valves—in a highly personal way, or a simply open or Harmon-muted horn. But more than anyone else, after Davis himself, I hear Louis Armstrong. There, I said it.

I said it for this reason: several people have tried to describe the emotion they hear in Miles Davis. It has been called effete lyricism, forceful lyricism, ecstasy. One man says that he hears in it nothing but defeat and despair. Another hears the whining of a disgruntled child. For me there is, beneath the sophistication and the thorough transmutation, the same kind of exuberant, humorous, committed, self-determined, and forceful joy in Miles Davis that there is in Louis Armstrong. *(1961)*

Errors of Interest

Jazz, I read in a recent issue of a national magazine, was born in the brothels of New Orleans. I thought that we had got over that fantasy but apparently not.

The publication in question was one of those men's mammary monthlies, and I suppose the editorship must have been delighted with the idea. There reportedly are not many births that take place in brothels; the idea that a music was one of them must have been quite intriguing.

It has long seemed to me that the unconscious function of female semi-nudity in the rabbit magazines is to convince us that the feminine form is not so enticing after all, but—with the help of odd lighting, and unreal colors in the printing ink—can be reduced to the same level as, say, Loretta Young fully clothed.

But I digress, and to come back to the main point, the idea that Buddy Bolden's band would ever have been allowed in a high-class New Orleans pleasure palace is either hilarious or depressing—or both, depending on one's point of view.

We decide where jazz was born, depending entirely on what we decide we mean by jazz. Or on when we decide that a particular American "folk" music became jazz. For example, if ragtime is pre-jazz, then jazz wasn't born in New Orleans, for ragtime was born elsewhere. If any kind of blues singing is jazz, then jazz wasn't born in New Orleans either.

However, no one would question that American music took a crucial step in the city of New Orleans and became a relatively complex instrumental music there.

The city produced Sidney Bechet, Jelly Roll Morton, the Dodds brothers, King Oliver, Jimmie Noone, Zutty Singleton, and Henry (Red) Allen—not to mention Louis Armstrong and to leave out several other very important musicians whom one should mention.

But jazz was, first of all, the music of certain communities in the city: the uptown Negro community, the downtown colored Creole community, and, quite soon thereafter, a segment of the white community. From these came the musicians and the audiences. They heard jazz in parades, at picnics, at dances, in bars, and elsewhere.

Once the music became established, it inevitably found its way into the local brothels. Countess Willie Pizzia, who as "the first lady of Storyville" ran one of those brothels, once told Kay C. Thompson, "Where jazz came from I can't rightly say, but . . . I was the first one in New Orleans to employ a jazz pianist in the red-light district. . . . In those days jazz was associated principally with dance halls and cabarets. . . . Jazz didn't start in sporting houses. . . . It was what most of our customers wanted to hear."

It was the pianists who played jazz in the brothels, then. (And incidentally, aside from brothels and bars, pianists seldom played jazz in New Orleans; hardly ever were they a part of the jazz bands in those days.) The pianists were Negro, colored Creole, and sometimes white. They were usually segregated, sometimes discreetly behind a screen, sometimes off at the side downstairs in the houses. (No, I wasn't there; I'm just repeating what I've heard.)

Jazz, like ragtime before it, was soon welcomed in brothels pretty much all over the country. And it was a part of the speakeasy-gangster scene in the '20s. This fact has led some writers to attribute a kind of simple-minded hedonism to the music. The opposition may well say that Brahms, after all, worked in a whorehouse, and Scarlatti performed his sonatas on a raised platform before the milling throngs at Italian fairs.

All of which may be irrelevant. Jazz and Brahms aren't important just because they were heard in brothels. But I do think that the association between jazz and the underworld in its various manifestations—which still goes on today, by the way—is a matter with complex psychological and social implications.

It would take a lot of exposition, thought, and research to find its meaning, I expect. But if I were going to search for it, I would start with this idea: jazz in some sense represents important aspects of American life, but aspects associated with unsolved problems, a lack of self-knowledge, all sorts of things (positive as well as negative) we don't admit or don't face up to.

It has to do with vital and crucial things about Americans that are not a part of the comfortable, innocent picture of ourselves we like to present to the world and to each other. Ironically, those unadmitted things are sometimes more joyful than the things we do admit to. They are also sometimes more painful, and they are always more tragic, which means that they are also more noble. That I think is why—or at least partly why—jazz is so compelling and so important.

At any rate, anyone who thinks jazz was born in a whorehouse ought to be made aware of just whom he is insulting: the uptown Negro community, the downtown colored Creole community, and a segment of the white community of New Orleans.

There are certain puzzlingly persistent errors in the literature and folklore of jazz that could stand a little airing from time to time. And, although I don't suppose that exposure will cause them to waste away and expire (having persisted this long, they may go on forever), it couldn't hurt to let some air get at a few of them.

For example, I keep reading, here and there, that modern jazz musicians began the business of making their solos out of chord changes rather than making variations on melodies themselves, that it was Charlie Parker and Dizzy Gillespie and their followers who began improvising on chords, inventing new melodies within a harmonic framework, rather than improvising on the themes themselves with ornaments and paraphrase.

I don't know where this idea came from, but it is emphatically untrue. All of the great players of the late 1930s invented on chords as a matter of course, night after night. Ben Webster did it, Roy Eldridge did it. Johnny Hodges did it, Buck Clayton did it, Benny Carter did it, Lionel Hampton did it, Coleman Hawkins. . . . All of them. Clearly Charlie Christian did it. And have you ever heard Teddy Wilson's solo on *Body and Soul* with the Benny Goodman Trio? It is one of the most brilliant harmonically oriented inventions I have ever heard in jazz, regardless of period. And it was no exception to Wilson's rule at the time.

Nor is such improvising limited to the great players of the late '30s. Almost every important player of the previous generation was at least capable of inventing on chords and did do it on some occasions. Louis Armstrong? Do you know the Victor version of *Sleepy Time Down South* or the last chorus of the Decca *I Can't Give You Anything But Love?* Jack Teagarden? Do you know his *Sheik of Araby* or his half-chorus on *Pennies from Heaven* from the Armstrong-Teagarden Town Hall concert on Victor?

There are many examples of Earl Hines doing it. And Red Allen. Sidney Bechet? Do you know his version of *Sleepy Time Down South?* Bunk Johnson? Yes, even Bunk Johnson. Have you ever heard the last two choruses of his

Some of These Days? And with Bix Beiderbecke, playing on chords was almost the rule.

Associated with that story is the one that says the modernists not only started playing on chords but also started the business of writing new themes to old chord outlines borrowed from standard tunes.

This is perhaps the most nonsensical story of all. By 1929 almost every big band had two or three pieces based on the *Tiger Rag* changes, one or two from, say, *I Ain't Got Nobody* or *After You've Gone.* And by 1932 *You're Driving Me Crazy,* a comparatively complex piece, had lost its melody and acquired the more jazzlike theme called *Moten Swing.* Almost every big band soon had one piece based on *King Porter Stomp*'s chord changes, and some had more than one.

At about the same time, *I Got Rhythm* had also lost its melody and became Sidney Bechet's *Shag,* the first of hundreds (thousands?) of new *I Got Rhythm* themes to come.

Try to imagine the Count Basie book of 1938 without changes borrowed from *Tea for Two, Honeysuckle Rose,* and *Diga Diga Do.* Or Ellington without *Rose Room, Exactly Like You,* or the ubiquitous *Tiger Rag.*

For that matter, dozens of ragtime chord sequences, simple as they are, showed up in the jazz repertory of the teens and '20s with slightly new melodies.

All this comes from the blues. Longer ago than we know, playing the blues could mean making spontaneous melodies within a harmonic framework, with no reference to a main melody or theme. By the late '20s jazz musicians had clearly begun to discover that they could also—so to speak—"play the blues" on chord changes that didn't belong to the blues. They spent the '30s exploring the idea. Similarly, they found they could also *write* new themes on old changes—themes that were more appropriate to their own idiom than the popular songs they came from.

Such fundamental practices did not change with modern jazz. What Parker and Gillespie and the rest did was find a new way of continuing and expanding the old ideas. Admittedly their harmonic, rhythmic, and melodic language was more sophisticated. But what they did was revitalize the tradition, not break radically from it.

By the way, it's possible for the fan or partisan of early jazz to delude himself when he insists that his Dixieland idols are always playing the melody one way or another. Often they are not. Often they are actually stringing together some 30-year-old Armstrong phrases to fit some fairly simple chords with no reference to a theme melody.

Technically, what probably happens here is more or less this: most Dixieland melodies are tied closely to their chords. And the players solo close to the chords too. So a listener often thinks he's hearing a melody he actually isn't.

Take *Tea for Two*—it isn't exactly a Dixieland piece, but it will serve as a good example because it is so familiar. If I play the chords of *Tea for Two,* I have virtually played the melody also through the leading tones in those chords. And if I pick my solo notes directly out of those chords or use only their lowest partials, I've suggested the melody, perhaps without playing it in any way.

Therefore, what our Dixieland fan is actually saying is, "I like what I'm used to."

Ah, and how many modern fans are actually saying the same thing in attacking John Coltrane or Ornette Coleman?

In 1939 I read in *Jazzmen* that the New Orleans Rhythm Kings lifted their success, *Tin Roof Blues,* from King Oliver's version of Richard M. Jones's *Jazzin' Babies Blues.* I have seen the story repeated many times since.

Having lived with both pieces on records for some twenty-plus years now, I confess that the only real similarity I hear between them is that they are both 12-bar blues.

Presumably the idea is their second themes are the same. They aren't. They aren't even that close. True, Georg Brunis's trombone solo on *Tin Roof* is apparently based, in its outline, on a tuba or trombone solo (sometimes it is also played on bowed bass) that often shows up in performances of *Jazzin' Babies* as well as in *Tin Roof*.

Very much closer to *Jazzin' Babies,* not only in the main theme but also in the trombone episode, are the closing choruses of a piece recorded in 1923 by Ollie Powers called *Play That Thing* and credited to Powers. There are other blood brothers and first cousins to *Jazzin' Babies* besides *Play That Thing,* but *Tin Roof* is a distant relative at best.

To make matters more interesting, Jones had a way of retitling his own pieces: the second strain of his *Riverside Blues* and the main theme of his *29th and Dearborn* turn out to be essentially the same. So, for all I know, *Jazzin' Babies Blues* may have several authorized titles too.

Maybe the mistaken notion came about through singers. I know that in her 1940 recording of *Jazzin' Babies,* Georgia White somehow used the *Tin Roof* melody. Maybe Alberta Hunter and Eva Taylor, who recorded the piece in the 1920s, did too (it is easier to sing).

While we're at it, there was a 1950s' song called *Make Love to Me?* Jo Stafford had a fairly successful record of it. It used the *Tin Roof* melody, properly credited to the New Orleans Rhythm Kings.

Going back a bit, have you ever heard it said that early bands, particularly New Orleans bands, always used banjo and tuba in place of the later string bass and guitar?

Well, of course I wasn't there, but I do know that to most

New Orleans musicians cornetist Buddy Bolden is the semimythical, semimystical father of jazz. The surviving photograph of Bolden's band, from about the turn of the century, shows a string bassist, a guitarist, cornetist Bolden, a valve trombonist, and two clarinetists. No piano, no drums.

It's possible the drummer wasn't there when the photograph was taken, but still there isn't any tuba or banjo. And this, mind you, is a shot taken outside a tent, on a job, not a studio photograph, for which the men might have picked up horns lying around the place just for a show.

There are photos of New Orleans groups from a few years later that show an occasional banjo or an occasional tuba or sousaphone, but they are not the rule.

Another idea that was popular, especially in the 1940s, was that the real New Orleans groups never used saxophones, the front line being composed of one or two trumpets, clarinet, and trombone. But the early photographs have a lot of saxophones visible, and the earliest recordings by several famous New Orleans players, including those by King Oliver and Jelly Roll Morton, have saxophones.

It is also bandied about that unison and harmonized passages were abhorrent to New Orleans musicians and they would play in the style of simultaneous improvisation, which used to be called New Orleans polyphony. (Strictly speaking, polyphony can mean several equally important melodies going at once, whereas in New Orleans jazz the trumpet melody is primary and those of the clarinet and trombone secondary.)

Except perhaps for the earliest records by the Original Dixieland Jazz Band, the music has plenty of passages in unison and harmony—only 10 minutes with early Oliver and Morton records will show this.

I would say that, on the contrary, one great virtue of the

best of the early jazz records is the great but uncluttered variety of effects these men used in a three-minute performance by seven or eight instruments: solo breaks and solo choruses, sometimes with rhythm, sometimes without; various combinations with two and three instruments in polyphony, unison, and harmony call-and-response patterns on a variety of rhythms.

Which is another way of saying that if you don't know Morton's *Black Bottom Stomp,* you should. *(1963/64)*

Roaring

"The Lion Roars" by pianist Willie "the Lion" Smith is part talk and part music. In the talk, the Lion is (understandably) not at all specific in his terminology or his categories. "Ragtime," "gutbucket," "lowdown," "in the alley"—they are all equivalent terms to him, as later are "ragtime," "jazz," "swing," "stomp," etc. And "rock and roll" comes from a description of the way gospels are done in Negro Baptist churches, not out of the brain of a disc jockey who didn't want to say "rhythm and blues." The historian and reviewer have got to be more specific about styles and movements, even if they seem arbitrary in doing so.

Ragtime was a thing "in the air" in both the East and Middle West before it found a maturity of form in Sedalia, Missouri, and the Eastern branch remained stylistically distinct, however, and later had a kind of center in Harlem. One can hear it in what must be an almost "pure" compositional form in Eubie Blake and Lucky Roberts. Later, as variation and improvisation began to become more pronounced and finally to dominate the style, it came to be called "stride" piano. The last of the great ones to retain the "stride" bass were, of course, Fats Waller and Art Tatum, but it is clear that the styles of, say, Count Basie and Thelonious Monk are later developments of the stride school. Traditional figures in the "jazz" (variational and improvisational) phase of stride piano are the late James P. Johnson and the Lion.

I think that if one compares their work with that of a similar transitional figure who made an improvisational jazz from ragtime materials, he learns a lot about both. Jelly Roll Morton did it earlier, elsewhere, and differently. For some reason which I have never been clear about, almost every stride pianist put Morton down. At any rate, in seems to me that his variations-on-theme were usually bolder, his rhythms freer, looser, and often more complex, and his invention more daring, except that Morton did not improvise noncompositional blues, and except perhaps in harmony. At his worst, an inferior stride pianist may sound, rhythmically, like some had wound him up. Morton's tempos might falter in his later years, but his time in shifting accents and introducing counterrhythms didn't. And very important is the kind of thing we see dramatized in the Lion's wonderful *Echo of Spring*. It is a leaning toward proper parlor piano, rather like Edward MacDowell's piano pieces. The tendency is present in some ragtime pieces, but jazz largely got rid of it. Meanwhile, a man like Eubie Blake had largely moved out of ragtime and jazz and into show music and society dance music.

A couple of details on the Lion himself and the record specifically: he is the most harmonically interesting of the "Harlem" school for me. This version of *Squeeze Me* is better than an earlier one (as *The Boy in the Boat)* and, like Thelonious Monk and Errol Garner, the Lion here shows some of the possibilities in a disciplined handling of a set of variations based directly on a melodic line. I hope that the fact that this *Fingerbustin'* is slower than an earlier version is not significant. Both *Bring on the Band* (after Basie) and *Portrait of the Duke* are excellent examples of *adapting* materials in tribute to another without copying them.

All of the narration the Lion offers on this LP, all of the arguments herein about who started what, when, and where boil down to this: both instrumentally and vocally, church music, blues, "ring shouts," jigs, cakewalks, blues, stomps, ragtime, and a near-ragtime style existed before, during, and after the New Orleans style grew and spread. But I don't know that anybody doubted that anyway. And I sometimes wonder if Lucky Roberts, James Johnson, and the Lion didn't all play pretty much the same way they would have played whether there had been any New Orleans jazz or not. But not Waller and not Ellington. (*1958*)

A Rock Cast in the Sea

"Don't you leave me here," sang Jelly Roll Morton. *"Don't you leave me here,"* he pleaded. But then:

> *If you just must go, sweet Babe,*
> *Leave a dime for beer.*

In the thirties, it used to be said that the blues were (in the phrase of the times) "social protest." Some of blues are, but comparatively few.

There are blues about man and nature:

> *Now the river she's gone to risin'*
> *And spreadin' all over the land.*

But most blues are about men dealing with other men, and sometimes, wryly, with himself:

I just got to holler, 'cause I'm too damn
mean to cry.

or:

I rolled and tumbled and I cried the whole
night long.
When I woke this morning, I didn't know
right from wrong.

And often with humor, even in despair:

I'm goin' down to the railroad
And lay my head on the track
When I hear that train comin' in
I'm gonna pull it right back.

But if you took a count, I'm sure you would find that the majority of blues are about courtship, with and without love:

Give a girl a dollar,
Next time you gotta give her five
Well, they ain't out for nothin'
Boys, but a line of jive.

The blues, on the whole, deal with universals—the things all men in all times experience and have to come to terms with; things that no man ever escapes, regardless of his circumstances. That is probably the highest tribute one could pay to them—higher still when we remember what grinding circumstances surrounded most of the blues singers as they conceived their songs.

Blues lore is full of striking, sometimes complex, poetic images:

The sun's gonna shine in my back door some
day.

(That is good, but I never knew quite how good until I asked myself why the *back* door.) There are also some wonderful word-plays and sound-plays, like this one on *morning, moaning,* and *mourning* by Sippie Wallace:

Early in the morning I rise like a
mourning dove,
Early in the morning I rise like a
mourning dove,
Moaning and singing about the man I love.

But beyond imagery, there is experience. This is one version of a traditional blues about a youngster's first love:

Early one Monday morning
I was on my way to school,
Early one morning
Goin' on my way to school,
On that morning
I broke my mother's rule.

And the same blues includes this wonderful traditional verse:

I was in love with you, Baby,
Before I learned to call your name.

Another blues climaxes with this fine irony:

Nobody on my mind
Carefree, sleepin' by myself.
The girl I love
Is sleepin' with somebody else.

(Carefree indeed!)

There are far too many blues about courtship and about sex perhaps—at least far too many for most of them to be very good. And inevitably many of them merely use the stock phrases to go through the emotional motions. Fewer still have any obvious continuity. You will notice that I have been quoting isolated images, lines, verses, many of them a part of traditional blues lore that every singer draws on, borrows from, and passes on to his successors. Frequently singers seem to string blues stanzas together with little regard to overall sense.

Sometimes they get the effect of great condensation and terseness this way; and sometimes the continuity is entirely emotional; but sometimes it just does not exist. There are cohesive exceptions. One that I am very fond of is *Fogyism,* written in the twenties for Ida Cox by her pianist Jessie Crump. The subject seems to be superstitions:

> *Why do people believe in some old sign?*
> *Why do people believe in some old sign?*
> *You hear a hoot owl holler, someone is*
> *surely dyin'.*

The next two verses go through several "signs"; the seven bad years from breaking a mirror; the curse of a black cat crossing your path; a dream of muddy water means trouble. Possibly, the song adds, it is the trouble that "Your man is sure to leave you." This leads the piece to a reversal and in the last stanza we are thrust from superstition abruptly into reality:

> *When your man comes home evil,*
> *tells you you are getting old,*
> *That's a true sign he's got someone*
> *else bakin' his jelly roll.*

Another exception, and surely one of the best blues poems, is Bessie Smith's *Young Woman's Blues.* It *is* a poem and almost as effective in print (it has been in a couple of anthologies of American poetry) as in performance. It begins:

Woke up this morning
When chickens were crowing for day,
Turned on the right side of my pillow,
My man had gone away.

Then she sings that she is still young, that she is a good woman (and anyway "Nobody knows my name/Nobody knows what I've done"). Through her protesting, we get a picture of her, of what her man was like, and what kind of woman he has probably gone off with. Now she is going out to have fun, "drink good moonshine and run these browns down." But the truth, as a lonely despair in her, increases to this terrible realization:

See that long, lonesome road?
Don't you know it's gonna end?

And, abruptly, a stoic recovery:

And I'm a young woman,
And I can get plenty of men!

In the beginning, music is song. And music, song, and poetry are all one. But from the blues of Blind Lemon Jefferson to the blues of Charlie Parker, there is, of course, great musical development. Poetically? Very little apparently. It seems there has been hardly any real poetic development in the blues since the twenties. And the best blues images are still a part of the traditional lore I have spoken of; the best singers repeat them. Even Muddy Waters's personal spell on *Hoochie Coochie Man:*

On the seventh hour,
Of the seventh day,
Of the seventh month,

The seven doctors say,
"He was born for good luck
and that you'll see."
I got seven hundred dollars
Don't you mess with me.

Surely this magic once belonged to a John Henry or a Stack O'Lee (regardless of the last two lines which I choose to take as an anti-climactic joke).

The one concerted effort by a poet to use the blues as a point of departure is Langston Hughes's, and his work remains a largely isolated phenomenon. Other poets have tried to use the blues on occasion: W. H. Auden, for example, but he has used only the blues verse form, not its personal directness, its ironic humor, its imagery, or its attitudes.

Some writers of more modest intentions still produce good blues, to be sure. Jesse Stone, for one (under his pseudonym "Charles Calhoun"), wrote a good blues song for Ray Charles called *Losing Hand* in which the love-game and the card-game exchange terms and images:

I gambled for your love, baby,
And held a losing hand.

And Jerry Leiber and Mike Stoller, the behind-the-scenes manipulators of "The Coasters" (among other rock and rollers) have supplied humorous songs in *Yackety Yack* (*"If you don't scrub the kitchen floor, you ain't gonna rock and roll no more . . . Don't talk back!"*) and in the one called *Charlie Brown,* about the classroom cut-up (*"Why is everybody always pickin' on me?"*) It is pretty light stuff, however, Creditable light stuff, but still light stuff.

There is one blues I have saved for the end, if only to show what might still be if blues poems had continued to develop from their roots as the music did—or perhaps it actually hints at something lost forever. It is Robert Johnson's incantation of inner spiritual torture, *Hellhound on My Trail:*

> *I got to keep moving, I got to keep moving.*
> *Blues falling down like hail, blues falling*
> *down like hail.*
> *I can't keep no money, hellhound on my trail,*
> *Hellhound on my trail, hellhound on my trail.*
> *I can tell the wind is running, leaves shaking*
> *on the tree, shakin' on the tree,*
> *I can tell the wind is running, leaves shakin'*
> *on the tree,*
> *All I need is my sweet woman, keep my company.*

Granted, the last line falters on a cliché, as too many blues do, although its idea is sound enough. But now go back to the beginning; the rest of *Hellhound* is powerful. And Johnson had other blues almost as good. (*1961*)

Mr. Wilson

Teddy Wilson can operate on two levels at once. To the casual listener his must sound like the pleasantest kind of unobtrusive doodling background piano. But, as in some of Mozart's "cafe" music or in Faulkner's *Saturday Evening Post* stories or in Shakespeare, there is a lot else being said than may show on the surface.

Nat Hentoff's very good notes, in effect an interview with pianist Dick Katz, are very clear about what the *else* is. There are probably half a dozen Goodman quartet records from the '30s which show it in retrospect: Hampton swings riff after riff with infectious energy; Goodman fingers his horn excellently, but in his quiet way, Wilson spontaneously invents original, cohesive, self-contained melodies.

"The Impeccable Mr. Wilson" is probably Wilson's best LP in quite a while, and it is performed by the version of the Wilson trio which was justly celebrated a year or so ago. Jo Jones and bassist Al Lucas work with him with sympathetic understanding of both the rhythmic and the emotional quality of his work.

However, a glance at the titles reveals what is probably one reason for the LP's success. They are: *I Want to Be Happy, Ain't Misbehavin', Honeysuckle Rose, Fine and Dandy, Sweet Lorraine, I've Found a New Baby, It's the Talk of the Town, Laura, Undecided, Time on My Hands, Who Cares?, Love Is Here to Stay. Here to Stay* and most of those tunes have been in Wilson's repertoire for years. He plays them excellently—even when he uses familiar lines (*Undecided* is a good example). He plays them with conviction, not as if he were giving some old material a run-through. But on some of them, say, *Laura, Time on My Hands, Love Is Here to Stay,* which we are apt to think of as possibly more recent additions, he does not do what he does best but essentially repeats the tune, embellishing it as he goes—drawing on his stock of pianist technique, I'd say, rather than on his imagination.

Even on those, the imaginative Wilson may assert himself—the second chorus of *Laura* is an example. And relatively conservative though they may be, I think that *Honeysuckle, Fine and Dandy,* and *Found a New Baby* are just about exemplary Wilson. *(1958)*

Navarro out of the Air

The selections on *"Fats Navarro with the Tadd Dameron Quintet"* are from radio broadcasts done from the Royal Roost in the summer and fall of 1948. As very special documents of a very special time in American jazz, they are invaluable. And they capture the excitement, the exuberant qualities of that time as no studio-made recordings do.

One can take two different positions about them as music, however. Kenny Clarke is himself, which is to say, one of the most smoothly infectious jazz drummers ever, a man with an important idea of the function of his instrument. But otherwise, one could say that what goes on here is a kind of pastiche drumming: a little of this, a little of that.

Tadd Dameron, as usual, plays "arranger's piano," solos in which not much happens except chords are stated and perhaps slightly embellished or a theme gets restated without really being reinterpreted. But wisely, his solos are brief, they have a kind of unapologetic movement to them, and they certainly do not bring things to a stall or a standstill. Rudy Williams, on alto saxophone, plays an unsettled, erratic cross between '30s alto and Charlie Parker.

Allan Eager offers early Lester Young phrases, one after another, and it is the most adroitly sophisticated pastiche imaginable. It is easy to resent what some players have done in borrowing Young's style, for some of them rattle off his ideas with an appalling disarray and illogic. But Eager—like, say, Wardell Gray—seemed to get more of the point, and, as he used Young's ideas, he captured some of Young's unexpected melodic logic too. At any rate, he plays with such youthful, dancing aplomb and such a clear sound that he is a delight, borrowings and all.

It also seems to me quite reasonable to say that, at his death, Navarro was still doing Dizzy Gillespie, with touches of Parker, Young, and others. Doing Gillespie in a personal way, to be sure, musically and emotionally, but still many a phrase, we know perfectly well, came from Gillespie. I readily confess, however, that I did not know how much of Fats Navarro, particularly his attack, was his own until I heard Clifford Brown after he had absorbed Navarro's style. Navarro did have his own way of doing it all, and he was a compelling trumpeter and improviser. I have played the bridge he uses in the first chorus of *Lady Be Good* (this is actually the *Rifftide-Hackensack* line rather than the melody itself) about fifteen times, and I still find it an almost unbelievable, breathlessly humorous episode. It is only one example.

If it were true that it ain't *ever* what you do but always is only how you do it, then Fats Navarro would have been one of the truly great players of his generation and Allan Eager more than a very good one. There is more to it than *just* how you do it, of course, but it was in how he did it that the really personal qualities of Navarro's playing lay. And they were very special.

This is no place to launch into a review of Dameron's career. He grew up writing swing arrangements and became "the disciple" of the modernists. I think some of his best lines from this period combine qualities of both styles, and I am glad that *The Squirrel* is included here, for it is a good example of the results up to that point. (It was originally an accompaniment figure on Billy Eckstine's version of Dameron's *Cool Breeze,* by the way.) The later Dameron, the Dameron of *Fountainebleau,* is another and bigger story. (*1962*)

How Long Has This Been Going On?

• I •

Wynton Marsalis is not the first jazz musician to perform classical music, but he is surely the most widely accepted and the most popular. Marsalis is also obviously one of the most musically outstanding young jazz musicians in over a decade. His work might be called a synthesis, a virtuoso summary of everything that jazz trumpet has ever achieved and ever been.

Marsalis is not the first about whom such a statement might be said. Almost a decade ago, it was obvious that Stanley Cowell (for one) was a comparable virtuoso for jazz piano and its history. Marsalis also has a counterpart in cornetist Warren Vaché, who in the past decade has explored the history of jazz cornet-trumpet tradition in public, and by now has learned bebop thoroughly. He has also explored the possible range of his instrument high and low, with commendable virtuosity.

All of which suggests that jazz is in a period of stylistic retrenchment or, if you will, a period of conservatism. And with Lester Bowie recording pieces that suggest nothing so much as the 1920s New Orleans style; with the presence of the O.T.B. group playing basically a late 1950s hard bop with all urgency and drive that almost suggests they invented the style; with the feasibility of jazz repertory orchestra on everybody's mind—with these and similar things—who could doubt that we are in a time of musical conservatism?

It is a conservatism of a sort that jazz has never experienced before, because, as is surely worth pointing out, the players and groups I have mentiond above, except for Vaché, are black, and that would mean that the most obviously outstanding young black players are (so far) doing nothing truly new. That has certainly never happened in jazz before: in the past, the kind of musical conservatism they represent has largely been the white man's burden.

To go back to Marsalis, however, and to be fair to him, there certainly is indication of growth in his work, and the signs are that the growth will come rhythmically—and that is something worth celebrating.

If I am correct about this state of affairs in the music, one might assume that jazz is in trouble, even that its future may be in jeopardy. I don't think so. Artistic retrenchment is not stagnation. It may be a necessary, even healthy state of affairs.

I have used words like *synthesis* and *summary* rather freely in the foregoing. If I applied the word *synthesis* to all jazz up to the 1960s, say, there would be a general understanding, a general agreement on what musical and stylistic elements, and whose accomplishments, were being brought together and synthesized—from New Orleans through early Monk, let's say. But for events from the mid-1960s onward there may not be a general knowledge of, or even a general agreement on, what went on, with whom, and with what results. So a bit of a review seems to be in order.

And if we are going to review the recent past, there is one phenomenon that seems to be standing in the way—fusion, jazz-rock, call it what you will. To go back to Wynton Marsalis, he has been quite outspoken on the subject—indeed, he, and some others, seem to see the whole fusion thing as a kind of commercial opportunism and artistic blind alley, maybe even a betrayal of the music, on the part of everyone involved, on the part of record companies, record producers, and the artists themselves. Marsalis wasn't the first to voice such disillusionments on fusion: several years ago, when he left CBS Records, Freddie Hubbard said some things which several observers had been feeling for some time, and they were the same sorts of things Marsalis has said. Nowadays, almost every month may find another musician voicing a disenchantment with fusion or quietly dropping the style.

Jazz-fusion (as though you didn't know) mostly finds its origins in Miles Davis's two-record album *"Bitches Brew,"* and *"Bitches Brew"* is indeed a witch's brew of jazz and every sort of then-popular pop style, black, white, and Latin, and the several participants in the various sessions that went into that album mostly moved into one or another of them as his own territory. Wayne Shorter, Joe Zawinul, and Weather Report chose jazz and electric rock. John McLaughlin and the Mahavishnu Orchestra, a version of rock that seems to have moved from acid trips to Eastern mysticism for its orientation. Herbie Hancock embraced a kind of jazz-r&b. Chick Corea, jazz and salsa. Keith Jarrett, on the other hand, mostly stayed on his own tracks.

Strictly on the fact of it, there is no reason why fusions might not have worked. Jazz has always been able to meet, absorb, and put to its own purposes almost everything in popular prospect—from the waltz and the tango through the works of the great popular songwriters and (for a brief moment) the *bossa nova.* And jazz has learned many lessons from the experience.

Furthermore, the idea of a contemporary fusion had already had its artistic successes. Thad Jones's jazz boogaloo, *Central Park North,* was surely one of his best pieces. Stanley Cowell's *Abscretions* is a dazzlingly cohesive piece made up of allusive fragments one could pick up listening to the local soul station. And with *In Front,* Keith Jarrett found something excellent in gospel impressions. (However Jarrett intended that piece, it comes off as a free-form impression of two kinds of gospel music.) Concurrently the Modern Jazz Quartet brought off a commendable jazz-reggae in *Walkin' Stomp* (did anybody notice?). For that matter, one might say that Ray Charles's band has (off and on) been playing a kind of fusion since the late 1950s. And I have a feeling that on a pop level Quincy Jones could fuse anything to anything with musicianship if he wanted to.

Nevertheless, I don't think jazz fusion worked very well, and I think my best way of explaining why not is to paraphrase something I heard Clark Terry say to a student band a few years ago: the beat in jazz moves forward; it is played so as to contribute to the all-but irresistible momentum of the music: jazz *goes* somewhere. The beat in most rock bobs and bounces away in one place—like the kids on the dance floor these days. Rock *stays* somewhere. And to be a bit technical about it, "jazz eights," the implied "triplet feel" of jazz, is rarely heard in fusion, and can seem strangely out of place when it is.

"Jazz" eight notes, the "jazz" triplet, are not the superficialities or the mere ornaments of a musical style; in jazz, they have always been among the fundamentals. One of the unwritten (and undiscussed) laws of jazz has been that each of the great players has found his own way of pronouncing the triplet, expressed or implied, and Roy Eldridge's triplet didn't sound like Louis Armstrong's; Miles Davis's didn't sound like Dizzy Gillespie's; Lester Young's triplet was unlike Coleman Hawkins's; and Stan Getz's is unlike Lester Young's. Nobody's triplet is exactly like anybody's. And developing a personally-articulated triplet not only has been an identifying mark for the great players, it has been an expression of the high individuality on which this music depends and which it celebrates. Also, swing is not simply a matter of musical momentum: that momentum is an aspect of the spontaneous, personal creativity which the music also celebrates. Swing encourages that creativity, makes it possible at the same time that it is an intricate part of it.

About Miles Davis and fusion, maybe I can be as blunt and outspoken as he usually is about everything and everybody. When I last heard Miles Davis he was stalking around a stage in what looked like a left-over Halloween fright-suit, emitting a scant handful of plaintive notes. A fast-fingered young tenor player, the piece being a B-flat drone, showed us how many ways he could run out of B-flat and back in again. A guitarist played with competence and little feeling. The whole thing reached the audience through speakers, each of which seemed bigger than the whole room I first heard Miles play in. And the music, like so much contemporary rock, seemed to get louder and louder in a desperate frustration over its inability to express anything.

Davis has always been one of those musicians who could come up with something so fresh, even on familiar material, as to make one forget, temporarily, all of his beautiful past. That evening everything I heard made me remember that beautiful past with pain. Soon I found that the booming of the speakers was producing a physical pain in my chest. I left, sad, and almost angry with the feeling that I had somehow been shut out by the music itself.

I am not going to tell you that all the efforts of a decade by such men as Wayne Shorter or Joe Zawinul, to name only two, have come to naught. There can be some nice houses even on a dead-end street. Those men and others are too talented as composers for that to have happened, and Shorter is too talented a player as well. But I do note that whenever I ask one of Weather Report's admirers to name a particularly favored piece, he will say, "Oh, something from (naming one of their LPs)." I can't imagine an admirer of Louis Armstrong, or Duke Ellington, or Charlie Parker, or an earlier Miles Davis reponding that way. He'd name you a piece or a solo, and then maybe tell you the LP it comes from.

• II •

The "New Thing" in jazz traces its origins to 1959 and Miles Davis's *"Kind of Blue"* album and Ornette Coleman's *"Shape of Jazz to Come."* But to say that is to leave out important events that preceded those albums, and it would be to leave out the work of a very important musician, Bill Evans. I have written on Evans before, but some of what I said should bear repeating here.

When he died in 1980, tributes to Evans came from pianists of several persuasions, but what most of his contemporaries and followers said about him had to do with the way he voiced his accompanying chords. Evan's harmonic language inspired Davis to call him a man who did not play chords but made sounds. That had something to do with his touch, of course, and with the flow of the music between his hands, one to the other. But it had even more to do with simple but crucial matters of harmonic technique. Bill Evans voiced his chords without using their root notes, and a chord without a stated root can have one of several identities: it can correctly accommodate a greater range of chords preceding and following, and most important of all perhaps, it can accommodate a greater choice and flow of melody notes and phrases above. Such chord voicings were fairly commonplace in classical music in the early decades of the century but they did not become so in jazz until the 1960s and after.

I don't want to leave comment about Evans on a more-or-less technical point, nor even at the assertion that he was a very influential musician. In any case, he surely lacked the slickness and glibness that I hear in some who have followed him who seem more affected by McCoy Tyner's fourths or by Ahmad Jamal's and Red Garland's practices on the matter than by Evans.

Evans could be the most fragile, the most private of musicians at times, but somehow since his death his very privacy and fragility have seemed to gain strengths of their own. At any rate, not to know the recordings of the Evans Trio with Scott La Faro and Paul Motian is not to know some of the treasures of recorded jazz. And not to know the posthumous *"Paris Sessions,"* with their new versions of *My Romance, Up with the Lark,* and *Nardis,* is not to know some revealing and exciting examples of continuing growth and development in a jazz musician.

Evans of course participated in the seminal Miles Davis *"Kind of Blue"* album, as he was superbly equipped to do, and in the two most influential pieces in that set, *So What* and *Flamenco Sketches.* Contrary to what you may have heard, neither one is really "free," yet in some respects, the later is freer than some things that have gone on since.

So What is a perfectly regular AABA song-form piece, and each of the eight-bar phrases is to be followed by the soloists as regularly as though the piece were *I Got Rhythm* or *Lady Be Good.* The difference (as is pretty well known by now) is that neither section has a chord progression. The A phrases have a mode—call it a scale—a succession of notes, which the player is to use in making up his melodies, and the B phrase has that same mode a half-step up.

Flamenco Sketches, by contrast, has no assigned phrases or phrase lengths. The player takes up a series of five modes in succession, improvising on each for as long as he pleases and then moves on to the next. So the basic shape and the length of the piece itself is improvised each time. In his solo, Evans improvises on the first mode (a C-Ionian) for eight bars, but in both of his solos Miles uses that opening only for four bars. And so on.

Obviously, the accompanists in *Flamenco Sketches* need to be alert when the soloists move from one mode to the next. And the way things have settled down in most modal playing, pianists will make their chords out of any selection of notes in the mode, but they will locate and include a stabilizing tonic note most of the time.

Bill Evans kept a few modal pieces in his own repertory throughout his career, beginning with *Peace Piece,* and he also had some pieces with irregular phrase lengths like *Blue in Green* and *Show-Type Tune.* But most of Bill's music was more traditionally oriented in variations improvised on the chord structures of popular songs, but chord structures voiced in new ways, and variations flowing in new directions.

John Coltrane was also a participant in the *"Kind of Blue"* music, and Coltrane, you might say, spent the next stage of his career exploring two directions at once. Pieces like *Count Down, Giant Steps,* and *Moment's Notice* call for an acute alertness to tightly structured, rapidly changing chord progressions, and they became the jazz etudes of the late '60s and '70s. Pieces like *Impressions* and *Love Supreme* are opposite: they are modal and therefore they have no chord changes, only assigned modes. Indeed, *Impressions* uses the same modes as *So What.* Only toward the end did Coltrane undertake a truly "free jazz," most notably in *Ascension.*

Charlie Parker spoke in 1947 of using the "upper partials" of chords as his melody notes. But the "upper partials" to Parker in 1945 were no longer "upper" fifteen years later. For John Coltrane, particularly after his experience with Thelonious Monk in 1957, upper partials were apt to be extensions like ninths, elevenths, and thirteenths. Not that such intervals had not been used in jazz—as enrichments in chord voicings, or as almost incidental "passing tones" in solo lines. But they had not been used as points of major melodic emphasis, and never so consistently as they are today. Typically, Coltrane would run up to these extensions, out of them, and on to the next, with scales (his "sheets of sound"). It was almost as if he were trying to show the way, to map out the new territory for everyone, to demonstrate their "correctness," step by step. (The man must have spent hours upon hours with Nicolas Slonimsky's scale books!)

• III •

Eric Dolphy's use of extensions was boldly selective. Necessarily, he formed his solos out of the same intervals, the same extensions which everyone else came to use, but on hearing him, some quite knowledgeable musicians said Eric couldn't follow a chord progression correctly, and was playing a lot of wrong notes. Why should notes that had been commonly used for years in solo melodies as "passing" tones, notes that for years had also been used to enrich chords, seem wrong in Eric's solos? Seem wrong, furthermore, to musicians, some of whom should have known better?

Perhaps because those intervals weren't used in passing in Eric's solos, and he didn't run up to them with scales. He used them consistently as his major melodic materials, and didn't resolve them in conventional ways. Perhaps most important, using a melody note that isn't actually stated in an accompanying chord is apt to sound strange to us, even though it could be quite proper to that chord—strange until we get used to it, anyway.

Of course such intervals used in such ways are commonplace nowadays. They are a necessary part of almost every young player's equipment. And this has to do with the real meaning of "playing outside," as musicians sometimes put it. Outside the tamer intervals, but not really outside the chords, not "free"—that is, not free of a chord progression.

During Eric Dolphy's all-too-short life—he was dead by 1964, a few days after his thirty-fifth birthday—he was inevitably compared to both John Coltrane and Ornette Coleman, but it seems to me that the comparisons might have been better made.

Charlie Mingus, with whom Eric made some of his most appealing recordings, once said that Eric had "absorbed Bird rhythmically." So he had, and so, I think, Coltrane had not. John Coltrane used a kind of heavy/light melodic rhythm which was basically as old as the very earliest New Orleans jazz, and which Coleman Hawkins and Don Byas had explored for years. It was almost as if, for Coltrane, Charlie Parker had not existed rhythmically, not to mention the rhythmic interest and variety in Louis Armstrong and Lester Young. And Coltrane—again like Hawkins and Byas—was primarily a vertical player, as I've indicated. Dolphy was a passionate, selective melodist. Like Armstrong, Young, and Parker.

Dolphy made his first recordings for a small label in 1948–49 as a member of a Los Angeles big band led by drummer Roy Porter (Art Farmer and Jimmy Knepper were also members), but those records are hard to come by, and I confess I have never heard them. Ten years later with Chico Hamilton, Dolphy had begun to sound rather like a Charlie Parker in search of something beyond. He was soon finding it, first on alto, almost simultaneously on flute, and finally on bass clarinet, the last of his three instruments to free itself from bop phrasing, from a bop melodic-rhythm.

Music was a continuing adventure to Eric Dolphy. Every

solo called for taking new risks and chances (and how many of us have the daring to do that with our own work and our own lives?). And he greeted music as he greeted life, with an inquisitiveness, openness, generosity, and enthusiasm that was as infectious as it was natural. As Gunther Schuller once put it, he knew that the more you learn, the more you realize there is to learn. But Eric was neither innocent nor naïve in his openness. He was alert, discriminating, selective—always with modesty and grace.

He worked with Ornette Coleman, of course, and he fit in with Ornette's music. Eric once indicated that Ornette's playing encouraged him in a direction, and I think that direction had to do with two things. Ornette's notes and phrases were more vocally inflected, more freely intoned than those of most earlier musicians. For Eric, it was a question of "getting the horn to more or less speak," as he once put it, not an easy task if one is at all serious and tasteful about it. (Some observers of course just had to decide that Ornette and Eric both weren't playing in tune!) Second, there was Ornette's "free" improvising, not only without any necessary sticking to bars and phrases lengths or forms but also without any chord progression, and without any *conscious* use of a mode or scale.

Ornette's solos do stray out of a basic key center, and in that sense can be called atonal. However, they can also be heard as modal. That is of course a kind of after-the-fact observation for a listener to make, however, and not, as I say, any part of Ornette's conscious intention. They just come out that way, you might say.

What Ornette did by an intuitive leap, Eric worked through step by step, from his work on chord changes, then heavily into their "outside" extensions, through the consciously modal improvisations everyone was learning in the 1960s, into a free improvising.

Likenesses aside, the results could be quite different. In his solos, Ornette seems to be running a kind of parallel creation to the tune, even though he is playing his own compositions. Eric, able to work inside, outside, modally, and free, sometimes seemed to be doing all those inside, outside, and free things at once, in solos that are melodically selective and direct, and in the voice and with the confidence of a man born to make and to communicate music.

• IV •

When I began all these words, I said that, although it may have produced some good music, the fusion effort seems to me largely over and was even something of a mistake. (Well, look, there can be some very handsome houses on a dead-end street.) And I said that I think that jazz is clearly in a conservative period of synthesis and retrenchment.

Musical conservatism and retrenchment are not stagnation, and past periods of conservatism in jazz have seen some of the music's major events. In the early 1930s the job at hand was for the bands to absorb and use the innovations of Louis Armstrong collectively, and at that period Duke Ellington produced such masterpieces as *Old Man Blues, Mood Indigo, Echoes of the Jungle, The Mystery Song,* and *Daybreak Express*. A decade later, when the task was to build on those Armstrong-inspired accomplishments, Ellington came up with *Ko-Ko, Harlem Air Shaft, Blue Serge*—a masterpiece almost every week, as Gunther Schuller once put it.

When Charlie Parker died, musicians began to look to Thelonious Monk, and they found many of the accomplishments of modern jazz synthesized in his music. At the same time, Charles Mingus, John Lewis, Gil Evans were taking their places as important composers in modern jazz.

If I'm right about all that, compositional synthesis seems

soon to follow on major innovation. To state that another way, a period of sound conservatism seems to follow on the arrival and absorption of something new. And to put the question of the current state of the music in something of the same way, is there an important composer for the jazz of the past two decades? Is there someone synthesizing the years of modal jazz, "outside" improvising, and free jazz?

The answer I think is yes, but that composer is not one man but four. And their compositions are a combination of writing and improvising rather unlike any that jazz has offered before, but in just the sort of synthesis the free jazz idiom calls for. I have in mind the World Saxophone Quartet.

Someone provides the bases for the WSQ's performances, of course, most frequently the group's alto, soprano, and flute player, Julius Hemphill, but each member has contributed to the group's repertory. The pieces, like bop heads, are vehicles, however, brought to life only as they are played and improvised on, and improvising for the group usually means in part a collective, multi-textured, polyphonic music that has become basic to free jazz. The Quartet brings it off masterfully, with an almost unbelievable lack of warming up, searching, and wasted effort.

They bring together a great deal, these four. Some of their pieces seem to be made up whole, as if the head melodies and textures, solos, all were improvised. And their polyphony is as basic and as uncompromising as it was in New Orleans seventy-five years ago.

I have written about the World Saxophone Quartet before, and I hope some of what I've said will bear repeating. Some people seem puzzled, even put off, by the fact that the group has dropped the rhythm section. But after all these years, and all the restless changes that the rhythm section has witnessed, we still have the same three men supporting one or two horns in much the same way as they do fourteen to sixteen horns. The Quartet has in effect cut the jazz ensemble back to its essentials, as Joseph Haydn did to the European orchestra when he arrived at the string quartet. The Quartet's members have done it for the same reason as Haydn: in order to produce some exceptional new music.

My view of the Quartet's place in the scheme of things, and my view of the course of jazz history before them—from major innovation to synthesis—is, of course, my own. I've offered that view before, more than once, and of course I need to admit that it may be a pattern of my own that I impose on jazz history. So it may also be true that I respect the music of the World Saxophone Quartet as much as I do because they seem to fulfill some expectation already had.

I do have reservations. I think that David Murray, more particularly in his own music outside the Quartet, tends to rush to his best effects and ideas, and that some of his work both as a writer and a player could use better pace, a better sense of structure—perhaps a kind of basic aesthetic *patience*—qualities which Sonny Rollins has so sublimely. Also, his *Bechet* seems to me a parody that borders on ridicule and not a tribute to that great musician. I would also guess that Hamiet Bluiett is also capable of ridiculing the jazz past when the mood is on him.

Julius Hemphill's piece *R&B* (Black Saint 0027) is a joy, including the in-studio conversation somebody decided to leave in the tape. And the sacred tributes—Bluiett's *Quinn Chapel AME Church* and Oliver Lake's *Hymn for the Year*—are heartening. But there should be much to come in everybody's work for the group.

(I also think it's the business of writers like me to say what we think, but I don't like the idea of giving advice to musicians, so please don't take the two paragraphs above that way.)

I suggest again that the best place to start with the World

Saxophone Quartet is the LP *"Steppin'"* (Black Saint BSR 0027), but I am in awe of the version of its title piece, *Steppin'*, on *"Live in Zurich"* (Black Saint BSR 0077), with Bluiett and Murray opening the piece a twelfth apart. On the other hand, their recent Ellington LP on Nonesuch, which seems to have sold very well, is largely homophonic, untypical, and to me not the best place to start with them.

The categories and labels that critics and historians set up to interpret events in the development of an art are never self-contained, and they do not (or should not) prescribe anything. When Ellington was doing all that superb work, he was also planting the seeds for change. And Roy Eldridge and Lester Young had been there for years before, reaching beyond the rest.

Which leads me to wonder: is the next innovator already out there? Is there a Louis Armstrong, a Charlie Parker, an Ornette Coleman already out there playing? If he is, when will someone notice?

This Fellow George Winston

As I remember, I first heard of pianist George Winston from a sign in a record shop window, "George Winston Back in Stock," while another shop two blocks down boasted "YES, WE HAVE WINDHAM HILL RECORDS!"

Soon after, an ad appeared for a local concert, and a clipping of a record review arrived through the mails which blamed George Winston (whom the reviewer obviously didn't like) on the influence of Bill Evans.

I decided I'd better hear this guy somehow, so I asked a friend at work if he had any George Winston records. Knowing my friend Tom's musical tastes, I figured he would have, and he did, and he generously made me a C-90 cassette covering two full Winston LPs. That's how I heard him.

About a week or so later I happened to be at a cocktail party (an unlikely place for me, by the way) when a middle-aged lady came up to me. I didn't know her, but she seemed to have a line on my musical interests, and she started raving away about George Winston. At a pause in the rush of enthusiasm, and without having given this proposition a moment's thought, I heard myself asking her, "Do you play the piano?"

"Oh, no!" she said quickly.

"Well, do you touch type?"

"Yes. I'm very good at it."

I cautioned, "I mean with all ten fingers, just the way you're supposed to?"

"Yes, and I'm quite fast with a little practice. Of course, it's been a while . . ."

"Well, you give me three hours"—I was trying to smile as I said this—"and I can teach you to do what George Winston does maybe just about as well as he does it, and I don't play piano *or* touch type."

"Really?" she said, dragging the word out at some length and looking shocked.

"Really."

It was the last I saw of her of course, but the next day I had a slight attack of remorse. "Oh boy, Martin, you've really done it this time: you've upset that well-intentioned lady no end, and you've probably been unfair to Winston." So I called up a couple of pianist friends and told them what I'd said and asked them if I had overdone it. The first one said quickly, "You haven't overdone it by much. Maybe not at all. Almost anybody could learn to do what he does. Come to think of it, maybe that's why people like him so much—they sense that they might be able to do it too."

The second pianist-friend not only confirmed what I had said, he launched into a technical explanation of exactly how almost anyone could learn to do it in very little time.

So, what about my friend Tom with the George Winston records? Well, I went back to him the next day, with his C-90 audio cassette in my hand. "Tom," I said, "you were very nice to do this, and I'm really grateful, but I want to give this back to you. Furthermore if you have any interest in any future relationship between you and me, I want you never to mention this incident again."

"Are you serious?" he said. He was grinning.

I said I was. Really. *(1987)*

II
APPRECIATIONS

In the 1960s, I contributed to a series of brief profiles which appeared in the International Musician, *the monthly publication of the American Federation of Musicians Local 802 (New York City). The occasion called for appreciations, not critiques or evaluations, and for an account on the state of the subjects' careers, their personalities, and their musical outlooks. I wrote of course from personal knowledge and personal observation, but it seemed appropriate also to draw on previously published material.*

Each of these men readily celebrates his early inspirations and influences, but each of them is also among the many distinct musical personalities and sensibilities without which jazz would be the poorer. And without which it could not exist at all.

Lee Konitz: A Career Renewed

In March of 1965, Carnegie Hall in New York was the scene of a memorial concert to jazzman Charlie Parker. The participants included trumpeters Dizzy Gillespie, Kenny Dorham, and Roy Eldridge; tenor saxophonist Coleman Hawkins; and, with Gillespie's group, tenor and alto saxophonist, and occasional flutist, James Moody. Also on Parker's own instrument was alto saxophonist Lee Konitz. Konitz says he wasn't sure exactly what he was going to do in his portion of the concert. When his time came, what he actually did do was step out on the large stage, alone and unaccompanied, and improvise a long and compelling *Blues for Bird,* still alone and unaccompanied.

For many members of the audience, the event was something of a homecoming, the first time they had heard Lee Konitz in several years. It was also something of a revelation, for Konitz's sound was different, somewhat harder, and his ideas were somewhat different, more assertive. Yet both the tone and the ideas were still identifiably those of Lee Konitz. Thus Konitz seemed more than ever the most original player on his instrument in the idiom of modern jazz after Parker himself. At the moment of the performance, Charlie Parker had been dead for ten years. Konitz seemed a musician reborn. Perhaps Lee Konitz will once again acquire a flock of followers and imitators.

Konitz's family was not musical, although two older brothers enjoyed a little household singing. Like many another American of his generation, Lee heard his first jazz on the radio, and by the time he was eleven, he was already involved in clarinet study—chiefly, he is sure, because of a youthful infatuation with Benny Goodman's music. He did not number among his young friends others who were budding instrumentalists and jazz enthusiasts, and Lee Konitz discovered recordings by master jazzmen like Coleman Hawkins, Johnny Hodges, and Benny Carter all by himself.

By the time he was fifteen, Konitz had worked in several dance bands around Chicago and even had his first experience on the road—this was during the manpower shortage of World War II, remember. He was already considered something of a hipster; he remembers wearing brown suede shoes with his tuxedo and doing a little improvising when he got the chance.

It was at this time, while working with a ballroom band led by Emil Flint, that Lee met Lennie Tristano; he was immediately taken with the blind pianist's work when he happened to hear him at another local ballroom. Tristano became the young clarinetist's mentor and teacher.

Within a few months, Konitz had taken up the alto saxophone, at first only because his jobs usually called for it. With Guy Claridge, he worked with guitarist Mary Osborne. With Teddy Powell, Konitz sat in the section with saxophonist Charlie Ventura. With Jerry Wald, Konitz was even more directly in contact with several aspiring young jazzmen.

Then came the first job which brought Konitz to national attention: in 1947, he joined the late Claude Thornhill and remained with him for ten months, during Thornhill's most important period.

It was perhaps foreordained that a young man with Lee Konitz's unique saxophone sound should find his way to Thornhill's orchestra. That sound has been called "legitimate." Actually it is neither more nor less like that of, say,

Marcel Mule than the sound of any other jazz saxophonist. But the Konitz sound, almost vibrato-bass, introspective in its effect, fit perfectly with the ideas of instrumentation and orchestration that Thornhill had been working on for some years.

The general style of the orchestra had been set by Thornhill himself and developed by arranger Gil Evans—the cool, harmonically rich sonorities, coming from the instrumentation which had trombones allied with the reeds, which used a French horn and a melodic tuba in the ensembles along with the muted brass, clarinets, and saxophones.

Partly through the encouragement of Gil Evans, the Thornhill band was also undertaking the repertory of modern jazz, with pieces like Charlie Parker's *Anthropology* and *Yardbird Suite* and Miles Davis's *Donna Lee*. In these scores, Lee Konitz's sound and approach, both as an ensemble player and a soloist, were obviously well suited. Something soon to be called "cool jazz" was obviously at hand.

On Thornhill's recordings of *Anthropology* and *Yardbird Suite,* Lee Konitz recorded his first solos. They are highly personal in style, and Konitz, who had come strongly under the influence of Lester Young's recordings while with Jerry Wald, was already well past his apprenticeship as a jazz improviser. At a time when hundreds of youthful saxophonists were simply repeating Young's ideas, Konitz had transmitted them and was forming his own approach.

In early 1948, Lee Konitz was in New York. For the first time he heard in person Charlie Parker, Coleman Hawkins, Roy Eldridge, Dizzy Gillespie, and he heard them with the excitement of an enthusiast who was, after all, still something of a youngster. Soon, because of his association with Gil Evans and Gerry Mulligan in the Thornhill orchestra, Konitz was asked to join the highly important nine-piece group of the time led by Miles Davis. The Davis Nontet was short-lived, and it spent much more time rehearsing the scores of Evans, Mulligan, Davis, John Lewis, and Johnny Carisi than playing them in public. But it recorded important and highly influential pieces such as *Israel, Boplicity, Jeru, Move, Godchild,* and the rest. Konitz's sound was important to the group, and Konitz was an important soloist. But today he feels that his participation in the Davis ensemble was hampered by his lack of experience at the time, and by what he insists was a still not fully-developed musical ear.

To have participated in such important activity as the Davis Nontet might be enough to give a musician mettlesome status in jazz. But at the same time, Lee Konitz was also involved in the appearance and the recordings of his ex-teacher, Lennie Tristano, particularly on highly regarded performances like *Marionette, Cross Current, Wow, Subconscious-Lee* (his own favorite), and the totally improvised experiment, *Intuition.*

In August of 1952, Konitz somewhat surprised the jazz world when, as a now-established soloist in small ensemble jazz, he became a member of Stan Kenton's orchestra. This was the Kenton ensemble which probably received more lasting critical praise than any of the leader's other groups. Besides Konitz, Kenton variously featured tenor saxophonists Zoot Sims and Richie Kamucha, trumpeters Maynard Ferguson, Conte Candoli, and Ernie Royal, trombonist Frank Rosolino, trombonist-composer-arranger Bill Russo, and drummer Stan Levy. Today, this ensemble is usually (and fondly) remembered as the Kenton "Young Blood" orchestra after a composition which Gerry Mulligan provided.

Between 1954 and 1961, Lee Konitz was either on his own or reunited with Tristano. As a leader, the saxophonist's career was begun auspiciously in Boston at the Storyville Club. He rejoined Tristano, often in the company of tenor saxophoist Warne Marsh, whenever the pianist decided to make a public appearance. Konitz also kept up his teaching, privately and, in the summer of 1959, at the School of Jazz, outside of Lenox, Massachusetts.

Konitz's modest musical philosophy of the time is indicated on the liner notes which he wrote for an LP called *"The Real Lee Konitz,"* which he edited from a full week's recording at a Pittsburgh club called the Midway Lounge. Konitz concludes his remarks, "All that I can say to conclude this short Apology is that five musicians gave all they had at this particular time in their lives, and will be grateful to know that someone might experience pleasure from their efforts."

In 1961 Lee Konitz moved to California. For over four years he lived in various places in the state, usually in the north. His reasons for the move were partly personal and partly, he frankly says, that he found himself a musician with a style somewhat out of fashion; New York jazz had gone "funky" and "hard." That, he told writer Ira Gitler, was "a big part of it," in his reasons for the move west. Sometimes, he took odd non-musical jobs—he painted bathrooms, he dug gardens.

Gradually however, Konitz tentatively re-entered music, first by doing some teaching in San José, then playing a little on weekends. But when he finally left California, he headed for his childhood home in Chicago, fully intending to enter the real estate business with other members of his family.

However, fate—once again in the form of Lennie Tristano—beckoned him back to New York for a job at the Half Note Café, and Konitz responded. Soon he knew that he was headed back into music in earnest.

After the engagement with Tristano, Lee Konitz continued the return, at first by a rather circuitous route. He began substituting in the orchestra at the Copacabana Club in New

York, and ended up doing two weeks there, partly accompanying singer Eydie Gorme—"it was fun," he says. At almost an opposite pole, Konitz, the improvising jazzman, found himself playing *a cappella,* both at the 1965 Newport Festival and the aforementioned Carnegie Hall concert. By mid-1965, he was off on a tour of Europe, one of the high points of which was an appearance at the (West) Berlin Jazz Festival.

Obliquely, Lee Konitz has also been involved in casual sessions and rehearsals with the New York jazz avant-garde, most fruitfully, he says, with pianist Paul Bley and guitarist Atilla Zollar. This activity should not be surprising. It was Konitz who was a chief contributor to the spontaneous, unpremeditated improvising with Tristano on pieces like *Intuition.* And almost eight years ago, Konitz made this statement about jazz: "I feel that in improvisation, the tune should serve as a vechicle for musical variations—and that the ultimate goal is to have as much freedom from the harmonic, melodic, and rhythmical restrictions of the tune as possible. . . ."

At the same time, he declares that, if it were somehow possible, he would love to play with Roy Eldridge and Louis Armstrong. *(1966)*

Lionel Hampton: Major Contributions

The vibraphone and the vibraharp are percussion instruments by virtue of the fact that they sound when they are struck. And they are apt to sit in the percussion sections of our symphony orchestras (when they are present at all) and be employed for occasional effect—rather like their wooden cousin, the xylophone.

Not so to our jazz musicians. To them the "vibes" in either version, as vibraphone or vibraharp, is a major melody instrument, and often it takes a front-line position in any ensemble in which it appears. It is capable of a wide range of sounds, and dynamics, and in the right hands it evokes a variety of moods. Our jazzmen have discovered resources in a number of instruments beyond the capacities previously conceived for them, of course, but perhaps none so much as the vibes. And the instigator and procreator of such discoveries on vibes is Lionel Hampton.

The story of how Hampton got started on the vibraharp, the version of the instrument he plays, is fairly well known by now. In 1930, he was a drummer with the Les Hite band in Los Angeles when Louis Armstrong was briefly fronting the group. And one afternoon during an Armstrong-Hite recording date, the young drummer began fooling around with a set of vibes which was sitting in the corner of the studio. Armstrong overheard him, and liked what he overheard. He used Hampton's vibes on his recording of *Memories of You* made that day, and he further encouraged Hampton to continue on the instrument. Hampton did, and a few years later he was a national figure as a vibist, known not only for his percussive medium and fast tempo showpieces like *Dinah* or *Sweet Sue* but also for his resourceful balladry on pieces like *The Man I Love, My Melancholy Baby,* and, once again, *Memories of You.*

National fame came about as a result of Lionel Hampton's membership in the Benny Goodman Quartet, of course, and it grew when Hampton subsequently went off on his own in 1940 and formed his own big band. Most followers of jazz think of this one as Hampton's first big band, but actually Hampton had left Hite and briefly gone off on his own as a band leader in the Los Angeles area, particularly at Sebastian's Cotton Club, before Goodman heard him and persuaded him to make the then-successful Goodman Trio into a quartet.

For the Hampton band, the going was a bit rough at first, but the ebullient 1941 Hampton recording of *Flyin' Home* established him as a colossal success. He remained leader of a big band, and he survived even the thin years for the bands, the times when some of the most famous leaders gave up altogether or cut back to smaller groups—something which Hampton has done only occasionally and more recently.

It seems impossible for any commentator on the scene to discuss the effects of Hampton's music without using words like "exuberance" and speaking of the joyous excitement with which he inspires not only his audiences but his fellow musicians as well. But Hampton frankly and proudly acknowledges, "Sometimes when I play jazz, it's like a spiritual impulse comes over me."

Many a musician, greeted with such adoring public adulation as Hampton has had, would be tempted to rest on his musical laurels and repeat himself with a kind of safe complacency. But not so Lionel Hampton. He deplores the stylistic cliques and fissures, musical and critical, which appear in the world of popular music, feeling strongly that there should be a fraternity among jazz artists of all persuasions and styles. The music, he says, has so much freedom and is so good to play. "Why do you decide jazz has to be a particular way?" he is apt to ask.

When bebop, the then-modern jazz of Dizzy Gillespie and Charlie Parker first appeared in the mid-'40s, Hampton welcomed it. And subsequently Hampton has been able to embrace and play his own version of any kind of music that people want to hear. When the mambo came along, Hampton played Hampton mambos. When rock and roll became popular, Hampton adopted some of its favorite devices. But none of this involves any compromise or dilution of his own ability. Indeed, Hampton has learned from subsequent styles, and expanded his own musical vocabulary over the years. "I can play anything I want on my vibes," he once told Bill Coss. Recently re-united with some of his former comrades of the swing era, he kidded them—in an entirely friendly way, to be sure—saying they should quit playing on those same old chord forms they'd been using thirty years and learn some new ones, the way he had.

Thus Hampton's inspired and inspiring exuberance, and his undoubted showmanship as a bandleader, is tempered by a knowledgeable and dedicated musicianship and discipline. George Simon, who has often been associated as producer with Hampton performances, wrote recently of a television rehearsal at which Hampton kept after his men, "making sure everything was going just right, and then refusing to quit when his allotted rehearsal time was over." The experience is not unique for anyone who has worked with Hamp.

Perhaps the basis of such musical tenacity and growth on Hampton's part dates back to his own early childhood and training—"Education in music is the finest thing that can happen to a child," he has said. He was born in Louisville, Kentucky, but his mother took him with her to Chicago when he was still small. Subsequently, he was sent to a Roman Catholic school in Wisconsin. There he was taught the rudiments of music and drums by a nun who was a strict task-master. His early experience was back in Chicago as a drummer with a newsboy band maintained by the paper *The Defender*. By 1928 he was in California and was soon a member of Hite's aggregation.

It would seem that Hampton has never lost touch with

the basic instruction he received in music, nor with the fact that it was a religious person who gave it to him. He is not himself a Catholic, but he has frequently helped Catholic, Jewish, and Protestant organizations of all kinds with fund-raising. This past fall, for example, he helped raise money for the Bishop Perry Hall in Harlem, a recreational establishment, and helped dedicate its opening.

This past summer, Hampton performed in the streets and parks of Cleveland and New York City to cool out the "long hot summer," and he is proud of the fact that, with the full approval of John V. Lindsay, he carries a card which reads "'Lionel Hampton, special representative of the Mayor," followed by the seal of the City of New York.

Early last November, he performed his *King David Suite,* written in 1954 and inspired by his first trip to Israel, with the Toronto Symphony, as he has with several major symphonic ensembles.

Hampton's enthusiasm for music is as genuine as it is constant. He often maintains the truth of the well-worn phrase that it is the "international language," by touring, and he has several times visited Europe, the Near East, and the Orient. Last November he set out for a six-week tour of the Orient with his current eight-piece group, the "Inner Circle of Jazz." On the agenda was Thailand and a visit (Hampton's second) with its king, who is a jazz fan and capable amateur musician.

As his performances in New York and Cleveland this past summer indicate, Hampton knows the human value of "live" music, and is always available to promote live jazz in any way he can, particularly to youngsters who have never had the chance to hear the music in person. At the same time, he understands the value of recordings to his career. He maintains his own record label, Glad Hamp, and also takes whatever free-lance recording assignments happen to interest him. Among the most successful of these, incidentally, have been a series of ballad recitals under such titles as "Golden Vibes" and "Silver Vibes." Perhaps such performances are a useful corrective also to those who think of Hampton only as the percussive enthusiast of *Flyin' Home* and *Hey-Ba-Ba-Re-Bop.*

Other bandleaders claim that they would never engage in a battle of music with the Hampton aggregation because "all he'd have to do would be to trot out 16 choruses of 'Flying Home'—so forget it!" Those leaders know that those 16 shouting choruses would contain a lot of good music. And that, for an encore, Hampton could easily pull off the impossible by offering his audience exceptional lyric explorations of some new standard like *What Kind of Fool Am I?* or *As Long As He Needs Me,* and make them like it.

Probably no musician is more concerned with reaching his audience, and Hampton's public statements are apt to be peppered with phrases like "getting to the people" and "putting the message across." "How about Leonard Bernstein?" he told Bill Coss. "I see him every time I can. I wish I had half the showmanship he has." Perhaps no man *needs* to be less concerned with rousing his listeners than Hampton, for showmanship comes as naturally to him as musicianship. And there is nothing at all false about either talent.

If he were tempted to look back, Hampton could look upon a career which includes besides his own achievements the encouragement of such arrangers and instrumentalists as Illinois Jacquet, Dexter Gordon, Earl Bostic, Benny Golson, Quincy Jones, Art Farmer, Joe Newman, Clifford Brown, Gigi Gryce, Kenny Dorham, Milt Bruckner, Charles Mingus—and the list goes on.

In his recently published book, *The Big Bands,* George Simon concludes his section on Hampton with some notes on character: ". . . he never hesitates to let anyone know he is grateful for any favor or kindness or other token of friendship. There is, so far as I have been able to determine, not an ounce of phoniness in this amazing man, who sometimes impresses you as a kid who has never quite grown up and then suddenly comes through as a remarkably mature human being . . ."

Perhaps what Simon describes in Hampton is the capacities of a mature being who has, at the same time, never lost touch with the child within. If so, the description is a fine tribute to Hampton the man. It is also, surely, a tribute to Hampton the improvising jazz musician. (*1968*)

Bud Freeman: The Needed Individual

In the spring of 1968, tenor saxophonist Lawrence "Bud" Freeman was announcing his full recovery from an automobile accident which had resulted in multiple fractures of the rib cage. His physician, who had at first warned Freeman that it might be more than a year before he could play again, had pronounced him fully recovered within six months. Playing the saxophone, Freeman reported to Jack Bradley in *Down Beat,* feels "better than ever. I feel freer . . . I've never practiced this much in my life."

One might say that the practicing paid off, whether he really needed it or not, because Bud Freeman is now involved in one of the most interesting projects of his career as a featured player with the Yank Lawson–Bob Haggart ensemble which has the marvelous name, "The World's Greatest Jazzband."

Freeman's continuing presence on the jazz scene is, or

should be, a reminder that there was a time in American music when there were just about two ways of playing jazz tenor saxophone, Coleman Hawkins's way or Bud Freeman's way. (Both men, however, pay respects to Prince Robinson as predecessor.) Even with the arrival of Lester Young in 1936, and the growing maturity of several of Hawkins's "pupils," there was still Bud Freeman and *his* progeny. That progeny is still there. And so is Freeman.

The story of Bud Freeman as a charter member of the Austin High School "gang" of young Chicagoans is a standard part of jazz literature. There were cornetist Jimmy McPartland and his guitarist brother Dick, bass-player Jim Lanigan, clarinetist Frank Teschemacher—these actually went to Austin. The other "Chicagoans," including drummer Dave Tough and pianist Joe Sullivan, did not.

It all began when a group of these youngsters happened to play an early 1922 recording by the New Orleans Rhythm Kings in a record shop. "I'll tell you," Jimmy McPartland has reported, "we went out of our minds. Everybody flipped. It was wonderful so we put others on. . . . We stayed there from about three in the afternoon until eight at night, just listening to those records one after another, over and over again."

Soon the group of high schoolers became, as Freeman put it to Ira Gitler, "a group of guys who would have nothing to do with anything but good jazz."

Tough, who had been taking professional jobs since he was about fifteen, was "the first to introduce me to jazz as played by the real players, and that was the old King Oliver Band." It featured a young Louis Armstrong on second coronet. Freeman and his young friends became, as Dick Hadlock puts it in *Jazz Masters of the Twenties,* self-conscious students of jazz for whom the music was "a challenging art that required deep thought and study."

Freeman, perhaps out of admiration for Jack Pettis of

the New Orleans Rhythm Kings, first took up C-melody saxophone, and the results by his own admission were not good. "I couldn't play anything. I could play one note." And McPartland, whose family was musical, has reported that "Bud Freeman was the only guy that had not had any training, consequently he was slow picking up the music . . . Tesch used to get disgusted with him and say, 'Let's throw that bum out.' But I said, 'No, no, no, don't. He's coming on, he's playing'."

The "coming on" was perhaps slow but it was sure. And it was built on a perceptive taste. "I was influenced by Louis, Beiderbecke, Bessie Smith, Ethel Waters, Buster Bailey, Fats Waller, James P. Johnson, Earl Hines—by drummers—Dave Tough . . . I was greatly influenced by jazz dancers."

By late 1928, a maturing Freeman, now on tenor sax and fresh from a Paris trip with Dave Tough, had made some recordings under his own name, the recently-reissued *Crazeology* and *Can't Help Lovin' Dat Man*. And by 1935, after several years of work with leaders like Roger Wolfe Kahn, Ben Pollack, and Red Nichols, But Freeman had become a featured member of Tommy Dorsey's band. "Tommy naturally had to feature what he did, which was a sweet, melodic trombone, but he did really have a great love for jazz . . . if a guy could play, he would really let him 'go.' "

The big bands, in Freeman's opinion, "needed individualists—they *needed* stars. Certainly, leaders might have had trouble with some of us, but we believed what we were doing, we grew up with jazz, felt strongly about our music and each of us developed in his own way, becoming both distinct individuals and *soloists*."

From Dorsey, Freedman went with Benny Goodman in 1938. And from Goodman, he returned to small group jazz with his own ensemble called the Summa Cum Laude Orchestra (in honor of the Austin High past), featuring Max

Kaminsky's cornet, Pee Wee Russell's clarinet, Eddie Condon's guitar, and Dave Tough's drums. This ensemble, in Hadlock's opinion, "developed into one of the most cohesive small bands of its time."

Then the entire unit quit night-club work to join *Swingin' the Dream,* a musical version of *A Midsummer Night's Dream,* and the show lasted only a couple of weeks after its Broadway opening. However, Freeman found more work for his group at home in Chicago, again in New York, and in an expanded, eighteen-piece version, on the road.

Back in Chicago for a rest and a visit, Freeman found himself with enough offers to keep himself busy, and he eventually ended up leading a kind of house band at the Sherman Hotel. "The owner . . . asked me if I could get a band in 17 hours. I called up eighteen men, rehearsed all night, and opened with a big review in the Panther Room."

Bud Freeman spent most of World War II as a member of the Army's Special Services in the Aleutians. On his return to civilian life, he worked for a while as house leader for the newly formed Majestic Record Company in New York. When the company failed, he returned to free-lancing.

In 1953, Freeman was back in New York after a tour of Chile and Peru, and a failed marriage. He was anxious to get to work again, but he had lost confidence in his playing. However, he was acquainted with some of modernist Lennie Tristano's recordings, thought him brilliant, and knew that he did some teaching. Freeman studied with Tristano for about three months, and, wisely, the pianist did not attempt to change Freeman's style but was able to help him re-learn his own way of playing. "He did give me terrific confidence," says the saxophonist. "He seemed to like what I was doing . . . I had to do what is me, what I honestly can say was my own playing."

Freeman then returned to the small groups which were

his first inspiration and first love. He became a member of George Wein's Newport All Stars, with time off for other projects, including a European tour. More recently he joined the remarkable aforementioned Lawson-Haggard World's Greatest Jazzband.

This energetic organization, which also boasts veterans like Billy Butterfield on trumpet, Lou McGarity and Carl Fontana on trombones, Bob Wilber on clarinet, and Ralph Sutton on piano, explains its billing by saying it is a *jazz* band whereas the other medium and large ensembles, with all due respect, are swing bands. Its repertoire, on the other hand, is up-to-date, and may feature Freeman on a quasi-Dixieland arrangement of, let us say, *Up, Up, and Away* or *Mrs. Robinson.*

Duly confident though he is, Bud Freeman still approaches each appearance with the kind of sound apprehension that a dedicated and sensitive improvising musician may be expected to show. On the road, he likes to reach his destination a day ahead and rest up, if possible. And before he performs, Freeman will probably complain that he doesn't feel well, has a cold, is tired, or whatever, and then go out and play his head off with the same genuine enthusiasm he has had since his twenties.

"I am interested in the individual," he says, returning to a favorite theme. "If he is sincere, I can see he's sincere." And, looking back, he added to Ira Gitler, "People responsible for jazz were individuals. If a musician believes in a thing, then the public will believe in it." *(1969)*

Thad Jones: A Musical Family

Thaddeus Joseph Jones is the musical middle brother of a Jones family which has more talent than any two, or any three, families have a right to. Older brother Hank Jones is one of the most well-equipped and dextrous pianists in jazz history (imagine a style in which both Art Tatum and Bud Powell are discernible as major influences). Younger brother Elvin Jones is an innovative drummer who, particularly for his work with the late John Coltrane, is considered *the* jazz durmmer of the 1960s by a whole generation of young players. Thad plays trumpet—or he did; nowadays, he is more likely to be heard on cornet.

Most musicians and fans, outside of the Detroit area, first heard Thad Jones when he joined the Count Basie orchestra in the mid-fifties. Jones, at thirty-one years old, was not then a young brass man still stylistically absorbing his youthful idols. He was already his own man, and to some he was a summary of the past of jazz trumpet which he had synthesized in a fully developed personal style, and to which had contributed enough brass techniques of his own to make him an original as well.

But Jones did not spring full grown from Zeus's head; he had been a working musician since his teens. He was born in Pontiac, Michigan, where his father, a lumber inspector for General Motors, had moved from Vicksburg, Mississippi. The elder Jones sang bass in his choir, and the oldest daughter of his large family, Olivia, although she died in an accident when only twelve, had already shown herself musical enough to be composing and giving piano lessons. Hank, Thad, Elvin, and the other Jones children grew up in the Depression, but as Elvin recently remarked to Whitney Balliett, "I guess we were lucky, because my father always worked. There was plenty of food even though I never saw any money." And of the Jones mother, he said, "She tried to make you into a man before anything else, so that you learned to take care of yourself, you learned how to survive. That was especially valuable to me in the beginning as a musician."

As indicated above, the three musical Jones sons were professionals by their late teens, working together as a trio during the late 1930s, and working separately—Thad did one job in Saginaw, for example, with a soon-to-be-famous young saxophonist named Sonny Stitt.

Between December 1943, and April 1946, Thad Jones was a member of the United States Army. On his discharge, the enterprising young trumpeter formed his own band in Oklahoma City. Within a couple of years, he was back home, working in Detroit with the group of tenor saxophonist Billy Mitchell at the Blue Bird. That small club and the Rouge Lounge became showplaces for young local musicians and young visitors, including pianists Barry Harris and Tommy Flanagan, bassists Paul Chambers and Doug Watkins, guitarist Kenny Burrell, and baritone saxophonist Pepper Adams. The visiting celebrities who were pleased to work with such local sidemen and accompanists might include Sonny Stitt, Milt Jackson (returning to his home town), Miles Davis, or the late Wardell Gray.

The legend spread about a "Detroit school" of young modern jazzmen, and in May of 1954, Thad Jones—recently off the road with a "Larry Steele Review"—got the aforementioned bid from Count Basie and joined his band.

Basie was well into his second career as a successful big

band leader (he had briefly cut back to a septet in 1950), with such attractions as Frank Wess and Frank Foster as his tenor sax soloists, Henry Coker and Benny Powell on trombones, Joe Wilder and Joe Newman on trumpets, and singer Joe Williams.

Basie also had a hit arrangement of *April in Paris,* and during it Thad Jones was assigned to interpolate a "Pop Goes the Weasel" lick at least once every night. However, Jones was not only given such exclusively comic assignments, and he had a particularly eventful solo on Frank Foster's *Shiny Stockings,* which is probably *the* Basie classic of the period. Thad also began to contribute to the Basie book, and, overall, he has offered *Mutt and Jeff* (for colleague Frank Wess's flute), the remarkable *Speaking of Sounds, The Deacon* (a blues of course), the ballad *For You,* and even *HRH* (*Her Royal Highness*), for Elizabeth II.

Jones, meanwhile, was recording on his own for bassist Charlie Mingus's Debut label—Mingus was an early Thad Jones champion—for Blue Note, and others. Now anyone who could not get enough of what Nat Hentoff has called Thad Jones's "satisfyingly brassful and singing tone" and his "individuality, maturity, and continuity of conception" had somewhere to turn.

In 1963, Thad Jones made an important move. He left Basie, but not to go off on his own as a leader. Rather, he more or less followed his brother, Hank, into studio work. He had all the qualities required: he reads a trumpet part, even a difficult one, instantly; he of course understands the jazz idiom (as most studio men must these days, since so much of the music they play is in that idiom), but he can play other idioms; and he can improvise and solo as required. He worked radio, television, recordings.

By November of 1964, Thad had joined Hank as a CBS-TV staff musician. And so he had achieved a measure of stability and security. But what he had not achieved is undoubtedly reflected in the fact that—to take some random examples—Thad Jones wrote an LP album of arrangements for Count Basie, *"Not Now I'll Tell You When"*; he wrote arrangements for the Harry James band; he worked as a featured soloist in two big concert bands built around the presence and the music of pianist Thelonious Monk; he worked with Gerry Mulligan's big band; he toured Europe in 1964 with the avant-garde group of composer-pianist-drummer George Russell; and he worked with his own quintet (featuring Pepper Adams) the following year.

Such activities represent a partial solution to a problem which is probably clear enough: how is a working musician who is a capable, quick professional to achieve full musical satisfaction if he is also an imaginative arranger and a creative jazz soloist?

In December of 1965, Jones and drummer Mel Lewis arrived at a solution, and it was a solution which has carried along almost a double-dozen of other musicians facing a similar dilemma. It started as the Jazz Orchestra, a cooperative big band of some eighteen players, a rehearsal band playing in an uptown Manhattan studio purely for the pleasure of the participants, as composer-arrangers, players, and soloists.

From the first, there was a feeling among all the participants that something very special was going on. There was a spirit of friendship and mutual respect, and there was also a high level of musical achievement. "It's fantastic," Jones told Dan Morgenstern soon after the project was under way, "I'm thinking about nothing else but this band. A lot of things had to be done; we had to get men who we felt were compatible, musically and personally. So far, it has worked exceptionally well. Everybody has respect for everybody else as musicians and people. Sometimes I get a little carried away hearing all this spirit coming through the horns." The band is "like a beautiful friendship that has been smoldering for a long time, and now it has burst open and is enveloping everything."

Co-leader Lewis has explained, "If you get guys who need work, you can't keep them. The guys in this band are working. The daytime is for commercial jobs—the night time is for jazz."

But the daytime, too, has on occasion become time for the Thad Jones-Mel Lewis Jazz Orchestra. First, disc jockey Alan Grant heard about the group as a rehearsal group. He took Max Gordon, owner of New York's Village Vanguard, to hear. The orchestra was booked into Gordon's club for four Monday nights; it has stayed on Monday nights since. It has also played concerts, clubs, and most of the major jazz festivals here and abroad, and has a recording contract. It is, to its avid admirers, one of the best big jazz bands there is, or ever was.

From the beginning, Thad Jones felt that the ensemble should "establish a style, a musical pattern. But it should have a lot of elasticity. Once you begin overlistening for something, what you're striving for is gone. And there has to be both freedom and discipline." Of his own composing, he knows that he has a degree of freedom with the group, "but I don't think you should use a band as a vehicle to exploit your writing. If something fits, okay, but if it doesn't, forget it."

The ensemble necessarily has something of a rotating personnel, but among the early regulars were Snooky Young, Jimmy Nottingham, Bill Berry, and Richard Williams on trumpets; Jack Raines, Tom McIntosh, and Cliff Heater on trombones; Jerome Richardson, Jerry Dodgion, Joe Farrell, Eddie Daniels, and Pepper Adams on saxophones; Sam Herman on guitar; Hank Jones on piano; and Richard Davis on bass.

Also one of its regulars from the beginning has been

Bob Brookmeyer, who writes for it, plays trombone in it, enthuses for it, and generally shows up for its every session unless he is deathly ill. Of its leader, Brookmeyer says that he has "all the components of a *man*—he's good, he's bad; he's sweet, he's dour; he's a leader of men and of music. Mostly, he's very, very good and a joy for another man to associate with. My life (and the world) would be diminished without his encouraging, benevolent, prodding, and human presence."

"He doesn't deserve verbal roses," Brookmeyer adds, with a flattering irony that is typical of him, "he just needs other men to live with as men were meant to live." (*1968*)

Bobby Hackett: Everything with Feeling

Trumpeter Bobby Hackett, in the phrase of a former associate, talks like a Boston gangster. Perhaps, but he plays like an angel.

Not that Hackett shares the vocabulary or the attitudes of a hood. And not that Hackett is a talker. And anyone who has tried to interview him will tell you that, when he talks at all, he would rather talk about others than about himself. He seldom misses the occasion of an interview to pay tribute to Louis Armstrong, for example. ("That man was and is the greatest hot-trumpet player in jazz. All you have to hear is four bars. None of us is one, two, three with Louis Armstrong.")

Praise for Hackett from his fellow musicians is consistently high. Pianist Dick Katz protests that he is probably saying only things that have been said before when he comments on Hackett's superb craftsmanship. "The performance standards he sets for himself are *very* high—sky high. His ballads are superb. He is a master of tasteful understatements. His sound is beautiful, and his clarity is marvelous—he never mumbles."

And Miles Davis, who is not one to hand out facile praise, has often praised Hackett and remarked recently, "A guy like Bobby Hackett plays what he plays with feeling, and you can put him into any kind of thing and he'll do it."

As Davis's statement might imply, Hackett wears a number of performer's hats these days. You may hear his graceful obbligato and shining middle half-chorus during the latest Tony Bennett performance, for example, or whenever Jackie Gleason decides to continue his series of mood music albums that began with *"Music for Lovers,"* Hackett will undoubtedly be there again, contributing graceful, uncloying middle choruses to piece after piece.

But if Hackett does everything with feeling, his first choice of things to do is to play in a club before a real, live audience as leader of his own small group. Even during his several years as a staff musician with the American Broadcasting Corporation in the late 1940s and early 1950s, he was periodically on leave to take club dates.

Robert Leo Hackett was born in Providence, Rhode Island, a belated Christmas present perhaps, on January 31, 1915, the sixth of nine children. His father was a blacksmith, the family was poor, and the gifted son was working professionally as a violinist and guitarist in the New England area, in restaurants and dance halls, by the time he was fourteen. He switched to cornet and trumpet when he heard Louis Armstrong—it was as simple and as deeply important as that. He began trying it out on the job. "I mean it," the manager of a Syracuse hotel is supposed to have said to bandleader Herbie Marsh, "if that guitar player picks up the cornet and tries to blow it just one more time, I'll cancel the rest of your band's engagement." It was perhaps the first, and probably the last time anyone ever complained about Hackett as a brassman.

By 1936, Hackett had taken over leadership of a group at the Theatrical Club in Boston, a group which featured clarinetist Pee Wee Russell and trombonist and arranger Brad Gowans, and he was attracting considerable attention. He came to New York in 1937 to play with Joe Marsala, then to play at Nick's in Greenwich Village, and thereby to become associated with the group of neo-dixielanders for whom guitarist and entrepreneur Eddie Condon provided leadership. In 1939, Hackett was on the road with a big band put together with the encouragement of MCA in its pre-television days. The venture was a failure with the public, but it left some recordings which were long in print and which have recently been reissued once again. One of them, a cleanly inventive version of Gershwin's *Embraceable You,* is probably the archetypal Hackett ballad performance, and it became an influential one among musicians. (Charlie Parker and the other "modernists" of the mid-1940s knew it well.)

Meanwhile, partly because of his essay, understated legato phrasing and his alliances with certain dixielanders, Hackett had become associated also with the Bix Beiderbecke tradition and was called on to re-create the latter's famous *I'm Coming Virginia* solo during the "historical" section of Benny Goodman's 1938 concert at Carnegie Hall.

Hackett worked in the tricky triple-tonguing trumpet section of Horace Heidt's orchestra in 1940 while paying off the debts of his own big band. ("I was the only musician that Heidt would allow to wear a mustache," he told Whitney Balliett, adding, "It was a mark of respect, I guess.")

The following year, a place was made for Hackett in the Glenn Miller band by having him dust off his guitar and play in the rhythm section. But he was given some moments on cornet too, and his solo on *A String of Pearls* has become almost a permanent part of that piece, usually re-created by trumpeters or even whole brass sections whenever it is played.

When Miller entered the wartime Air Corps, Hackett joined the NBC staff for a while, toured with the dance troupe of Katherine Dunham, and between 1944 and 1946 was a member of the Glen Gray orchestra. Then followed the off-and-on years on staff at ABC and the appearances on his own.

One of his moments-on-leave from ABC involved a historic 1947 New York Town Hall concert with Louis Armstrong, in which the latter cut back to a small group, and launched himself on the most successful period of his career. Another associate at that concert was pianist and arranger Dick Carey, a man with a tasteful and knowledgeable love of all styles and periods of jazz, and all kinds of jazzmen.

In 1957–58, Hackett led a remarkable sextet which used Carey's arrangements and which drew on all the then-existing jazz styles, New Orleans through Cool, and even including touches of Monk. The ensemble had an extended stay at the Henry Hudson Hotel in New York and made it to the jazz festivals.

During the past decade, Hackett has continued to work on his own, with groups ranging from trios to quintets. But last September he began an association which particularly pleases him, one he wanted for a long time, with trombonist Vic Dickinson. "Nobody . . . could say enough about Vic," he told Paul McKenna Davis in Toronto. "He does things on trombone nobody I can remember has ever done; he phrases out of this world. He never does the same thing twice. Every night, every session we play, he shows me something new—and beautiful." The two men are nearly perfect foils, and so far their press notices have been nearly ecstatic.

One of the members of Hackett's Henry Hudson sextet was Bob Wilber, who added some vibes to his clarinet for the occasion.

Clarinetist and soprano saxophonist Bob Wilber has worked with Bobby Hackett off and on as sideman and arranger for over twenty years, and has found himself constantly inspired by the experience. "His musical standards are so high that to play on his level is always a challenge. He is a perfectionist, always plays in tune, never misses a change, never stumbles rhythmically. Such utter consistency could get dull after a while, but Bobby is able to come up with real surprises often enough so that listening to him is always stimulating, sometimes exciting, occasionally thrilling."

Such surprises, Wilber continues, are particularly difficult "when you limit yourself, as Bobby does, to the basic chords, using very little chromatic alteration and also play very simply rhythmically (no polyrhythms, playing behind the beat, ahead of the beat).

"Of course, his secret weapons are that fantastic sound—the greatest legato of any player I've ever heard—and the beautiful use of vibrato. . . . And anybody who can find so *many* things to do with simple four-note chords . . . !

"But basically Bobby is just a fine man with a great ability to communicate with *everybody* through his horn. And the message is always love." (*1968*)

Harry Carney: Forty-one Years at Home

"My greatest kick with the instrument, which then seemed much bigger than me, was that I was able to fill it, and make some noise with it. I enjoyed the tone of it and I started to give it some serious study. I've been carrying it around ever since."

Thus, speaking to Stanley Dance, did Harry Howell Carney describe his first encounter with the baritone sax. Carney is, to most musicians and fans, the archetypal jazz baritone saxophonist, but it happens that he was well established on alto sax and clarinet before he took up the instrument. Actually, people seriously ask him if he invented the larger horn, but he was not the "first" jazz baritone man—nor the last, to be sure.

Carney's onetime Duke Ellington colleague, the late cornetist Rex Stewart, once said that "as a general rule, when an instrumentalist really makes it big, everybody tries to imitate him. However, Carney's concept is unique, so personalized that no one has been able successfully to copy his style or his famous sonority on the baritone saxophone."

Carney has, Stewart continued, "a range . . . that surpasses credibility. He plays the higher octaves not only in tune but with a tenderness that is sometimes mindful of a cello's sound. Then, when called for, he's able to attack the sonorous bottom of the horn with such vigor and vitality that it is small wonder he has been the anchor man in Ellington's orchestra for these many years." He has also learned a trick or two of breath control that enables him to hold the long tones that Ellington favors.

"There may be a poll he hasn't won," Stewart added, "but I don't know which it could be."

Neither of Carney's parents was particularly musical, but his own music education began in his hometown of Boston, with piano lessons in 1916 when he was six years of age. It did not go well in a sense; while Carney dutifully studied his classics, the neighborhood seemed full of self-taught little prodigies who could pick out popular ditties and the blues, and even ad lib on them, all on their own. To add to all this, his younger brother sat down on the piano bench one day and, with no preparation or study whatever simply started to play—at least, so Harry Carney remembers it.

Nevertheless, Harry kept up his scales and practiced his pieces for annual recitals, he went to school, and he sold Boston newspapers in whatever spare time was left.

Carney's was a musical neighborhood. There was Leonard Withers, a young pianist. There was James "Buster" Tolliver, who played reeds and later became a well-known arranger. There was Johnny Hodges, who moved across the river from Cambridge when Carney was a teenager. By that time, Carney himself had switched to clarinet. His first reasons were not exactly musical. He noticed that at dances, Tolliver played piano the first half and clarinet after intermission, and "he always seemed to be surrounded by the girls when he got through playing the clarinet, and by now I had reached an age when I was conscious of the girls, so . . ."

Tolliver explained that Carney could get clarinet lessons at a nominal fee, with the instrument supplied, if he joined the Knights of Pythias Band. Soon, inspired by Buster Bailey and Don Murray, he was, as he puts it, "alarming the whole neighborhood with my practicing," until someone thought he was good and offered him local jobs. "I was so anxious to prove to everybody I could play," he once explained to

Dom Cerulli in *Down Beat,* "that I'd just open the window and play loud. People used to say I slept with that clarinet. The truth is, I was never without it."

Soon he added alto saxophone, and found it easier to get a better sound on the large instrument. He and Hodges began to practice together. Carney was listening to Sidney Bechet's records by now, and Coleman Hawkins's ("Hawk was actually my idol . . . and he still is").

In the spring of 1927, when he was seventeen, Carney persuaded his mother to let him take a trip to New York with another soon-to-be-well-known alto player from Boston, Charlie Holmes. They went to see Hodges, now with the Chick Webb band at the Savoy, and before long Carney had a job or two himself. One of these, with Henry Sapro at a place called the Bamboo Inn, lasted a while. One man who heard him there was a young pianist-band leader, then on the way up, named Duke Ellington.

These were exciting days for the young instrumentalist. "I couldn't believe it," he told Cerulli. "I could see my favorite musicians every afternoon. Just to have a chance to talk with them meant so much to me. I used to eat at a restaurant at 131st Street. But this is how I ate: I'd order, and then run outside for a bit, then I'd come back in and eat a little, then I'd go outside again. I just didn't want to miss anything."

Then one day on the street, Ellington approached him: a New England tour was coming up, and Otto "Toby" Hardwicke wouldn't be making it. Could Carney come along? He could indeed. The trip, which would include Boston, might even help his homesickness.

The job with Ellington was supposed to have been temporary but it became permanent, which means that Carney has been with Ellington for forty-one years now.

There seems to be a minor disagreement among several

observers as to whether it was originally Ellington's idea or Carney's that he switch from alto to baritone (perhaps neither of the principals involved will say for sure by now), but Ellington certainly wanted the change and Carney certainly took to it. He acknowledges the influences of Hardwicke and Joe Garland (who played some baritone), but he told Stanley Dance that chiefly he "tried to make the upper register sound like Coleman Hawkins and the lower register like Adrian Rollini," whose bass saxophone he admired.

The first important job the Ellington band had after Carney joined was at the Harlem Cotton Club, and with the club's success and with nightly broadcasts, Ellington's career as a major figure in American music was under way. The subsequent story is one of growth—for Ellington as composer and leader, for his ensembles, and for Harry Carney on the instrument of his final choice. "Fortunately or unfortunately," he explained modestly to Dance, "there's nearly always been a better clarinetist in the band. I left the clarinet up to him."

As anyone who has watched Carney for only a few minutes will conclude, *modesty* is a key word. "If it is true," Stewart suggested, "that early environment shapes the individual, as I happen to believe, then it becomes clear why Carney developed into such a likable human being, since he is the product of a most harmonious household. I well remember how his parents always extended themselves in making Harry's band-buddies welcome every time we played Boston."

Carney still calls joining Ellington his greatest single achievement. "I was always surprised when fellows left the band," he says. With Ellington, "there was always something going on. I just loved it. He was always experimenting. And I liked his outlook. I liked the way he thought about music. It was right up my alley."

The mutual respect took a somewhat different turn about 1949. Carney, whose hobby up to that time had been photography, indulged his longstanding admiration of cars and bought one. He soon acquired a regular passenger on band hops: his boss. "Duke sleeps occasionally but not as a rule. He's a very good man to have along." Ellington thinks, or he makes notes. "We do very little talking, but if he thinks I'm getting weary, he'll make conversation so that I don't fall asleep."

Being on the road has its good side. "There's one thing about traveling: it always gives you something to look forward to, even if it's no more than going to another town to see people there you know."

Carney remembers those people, and they remember him. Stewart recalled that, "when the mood strikes him, he will get on the phone and spend hours calling all over the country to his friends. He carries several little address books with him, and his friends can expect to hear from Harry some time during the year." And when the orchestra takes an intermission, it is Carney who willingly chats with the people and signs autographs. "I have often watched him snatch his horn from his mouth when he had a two-bar rest to inform someone of the title of the tune the band was playing . . . while other musicians . . . ignore the question."

Carney was once a fan himself, and he obviously remembers how it felt. He also reads the jazz press, American and foreign, with much the same eagerness as he had as a younger man. The young fans often seem more serious in their questions to him than they once did—they ask about reeds and mouthpieces, about breathing. "But I feel jazz has become a part of American culture today. It's a language that is spoken everywhere. And it's not only the music, it's the people who play it."

Carney has been playing it for over four decades as a

leading instrumentalist with a leading orchestra, and as is often remarked, he looks almost as young as he did thirty years ago. "I liked the night life and the people we'd meet," he says looking back. "I looked forward to going to work every night. And I still do." (*1969*)

III
ON THE JOB

Condition Red

Trumpeter Henry (Red) Allen, Jr., has been recording as leader of his own groups since 1929, but, like many a veteran professional, he still approaches record dates with a bit of apprehension and a slightly nervous determination that everything shall go well. At least he did have such apprehension when he was to do a date for the Prestige/Swingville label recently, using the quartet he has been working with in clubs like the Embers in New York City and the London House in Chicago.

The session had been set up by Prestige's Esmond Edwards for 1:00 p.m. in the New Jersey studios of Rudy Van Gelder, across the George Washington Bridge from Manhattan Island.

Red Allen, with his group, pulled up in his car in front of Van Gelder's forty-five minutes early. He wanted everything to be relaxed and easy. Van Gelder—more used to lateness than earliness—was surprised and a bit dismayed by the arrival. But with a firm reminder that the date would not begin until one, he opened his door to the quartet. "Early—this group is always early," said drummer Jerry Potter, with a half-smile that didn't exactly reveal his feelings on the matter.

The day itself had held little promise as a day. The sky was overcast, there was a drizzle, and by late afternoon, when the date had ended, a heavy rain was falling. But inside the high-ceilinged, wooden-beamed studio there was plenty of time to set up the drums, plenty of time to get acquainted with the room, and even time for Allen to go over his lyrics and review the list of tunes he wanted to make. He leaned over on the back of the studio piano and scanned his papers, wearing a pair of glasses that gave him a studied air, an air that few who have watched the exuberantly powerful Red Allen on the bandstand would recognize.

As the men waited, there was a casual exchange at the piano bench. Not once did the group's pianist, Lannie Scott, sit down to noodle. It was the bassist, Frank Scaate, who played first, and later Allen. Musicians take this sort of thing for granted—nearly everyone plays *some* piano and enjoys it—but it is frequently surprising to outsiders.

A little before one o'clock Edwards arrived, also a bit surprised that the group was fully assembled. He took his place inside Van Gelder's booth, behind the large glass panel which is broad and high enough to take in the whole barnlike studio at a glance, and laid out his note paper and recording data sheets. Van Gelder soon had his machines threaded with tape and was seated behind his complex control panel. The date was officially ready to begin.

On the other side of the glass, the musicians began running through the first piece, *Cherry,* to warm up and to check the placement of the microphones. Allen was swinging from the first bar, and his very personal, often complex, phrases rolled out of his horn with an apparently casual ease. He was showing his fine control of the horn too. He would begin with an idea at a mere whisper of trumpet sound and develop it to a powerful shout at the end of his phrase—the kind of dynamics that few trumpeters employ.

After the run-through, *Cherry* was ready to go onto the tape. Take 1 had an inventive opening by Allen, but he stopped after his vocal, saying, "I goofed the words all up." Another take, but the bass wasn't balanced. First numbers on a record date usually go that way.

Then—*Cherry* No. 3. Everyone was working, and the

group was concertedly alive. Allen was truly inventive, for he used only one brief phrase that he had played in any previous version of *Cherry* that day.

"That man really improvises," someone in the booth said. Edwards and Van Gelder nodded agreement. "I wonder if he could repeat himself, even if he wanted to?"

As the ending rang out through the wooden rafters and across the mikes, warmly echoing the power and drive of the performance, Edwards was laughing and saying, "They don't play like that any more!"

"Can we hear that back?" Allen asked at the end.

A bit later they began running through *Sleepy Time Gal.* Allen's lines were weaving in unexpected but logical directions, and he was beginning to show his command of the full range of his horn, with the perfectly played low notes that are almost his exclusive property. His melodies were still gliding over the rhythm section and the time with sureness and inner drive and no excess notes.

The first take of *Sleepy Time Gal* was much simpler than the run-through, and there was some trouble with the introduction. Allen is still more used to recording for the flat acetate record blanks than for the more recent magnetic tape, and he had been counting off the tempos to the group at a whisper. But with tape it's easy to remove a spoken count-off. "You can count it off out loud, Red," Edwards reminded him.

At the end of another take, Edwards apparently saw something was about to happen, and he reached for his mike to ask over the studio loud-speakers, "How are the chops? Can we do one more right away?"

"Yeah, sure, my man!" Allen said immediately. And then they did the best *Sleepy Time Gal* yet.

This time Allen came into the engineering booth to hear the playback and sat beside Van Gelder's elaborate array of dials and knobs. He raised and curved his eyebrows at a particularly lyric turn of phrase in his own improvising, pretty much the way any listener would in following the music.

By 2:00 p.m. they were into *I Ain't Got Nobody,* and on his vocal Allen was getting in as many as six notes just singing the word "I."

After the run-through, Edwards suggested Allen blow another trumpet chorus on the final take. Again, Allen's ideas were fresh and different each time they ran the piece down, and he still glided over the time with perfect poise. His trumpet alone might make the whole group swing. He counted them off loudly now for the final take: "One! Two!" And at the end, after the reverberations had settled, there was the inevitable Red Allen genial cry, "Nice!"

Then a short break as visitors arrived. Van Gelder immediately gave them a firm invitation to sit quietly in the studio and stay out of the booth. Drummer Potter came in to ask for a little more mike on his bass drum: "Can you bring it up a little? Then I can relax. I have to keep leaning on it otherwise. Like playing in a noisy club."

"Okay, we'll try," Van Gelder said. "It's not easy to do."

In the studio, a photographer, there to get a shot for the album cover, had his lights and shutters going. Allen wasn't bothered. Nervous or not, he had been taking care of business from the beginning, and he was obviously impatient to get back to work.

Later, they were well into *There's a House in Harlem,* with Allen getting deep growl effects on his horn without a plunger. Again, every version was different. Van Gelder remarked for about the third time that they should be recording everything, including the warm-ups and run-throughs, and again shook his head in appreciation of how well Allen was playing.

Edwards stopped the take, remarking on the intro, and

pianist Scott and bassist Scaate worked it out together before the tape rolled again.

They began *Just in Time*. "Everybody plays that thing now," a visitor remarked. "I guess it's become a jazz standard already. I heard Art Farmer do it the other day."

There was some trouble again with the intro so Allen took it himself, unaccompanied. They went through the piece once, and Allen was after Potter: "Let me hear a little more of that bass drum, please."

Another break. This one was officially called by Edwards, who was concerned and impatient about Scott's intonation. Allen was still eager to get back to work, and he toyed around on his horn with the next piece he wanted to do, *Nice Work If You Can Get It*.

"Johnny Hodges has a record of that," remarked Scott. "Did you hear it?"

A bit later, when Edwards suggested they go back to work, Allen had relaxed at least long enough to be showing a visitor a color picture he has of his mother, himself, and his granddaughter—four generations of the Allen family. But he broke off abruptly and went back to his mike.

On the take of *Nice Work,* piano and bass took it partly in "two" (ah there, Miles Davis). "Make it clean." Edwards had encouraged them during the run-through. Allen's variations rolled off easily and with a rare and personal symmetry.

The quartet then began to run down a piece that seemed both familiar and not familiar, a piece that sounded like the blues and was not exactly the blues, and 32 bars. When they got the routine set, Edwards asked for the title. *Biffly Blues,* said Allen—so it was a new version of the first record he ever did under his own name. One take, and for the time being everyone agreed with Edwards's comment, "That's it. It won't go down any better than that."

As they were running through *St. Louis Blues,* there was talk in the booth about "still another record of that one." But Edwards decided that if they did something different with it, then it should be recorded. They did.

It was getting late, nearly 4:00 p.m., and Edwards did some quick calculations from the timings recorded in his notes on the session.

"Red, why not stretch out with a few more choruses on this," he said into the studio mike. "We'll have enough time for it on the LP."

While the tapes were rolling, Allen suddenly played very low on his horn again, growling out notes for almost two choruses. One take—as usual—did the blues.

The date was nearly over now. Edwards made more calculations on timing, and then stepped into the studio to suggest to Allen they do a longer version of *Biffly Blues.* Agreed.

"What does that title mean, Red?" a visitor asked hurriedly, hoping to get his question in before the tapes rolled again. "My nickname—when I was a kid," he smiled. "My folks used to call me Biffly when I wanted to be a baseball player. You know—biff—hit. Wham!"

After a rough start, occurring because Allen had placed his horn and set his chops too quickly, they got through a long taping of *Biffly Blues,* with Edwards conducting and encouraging through the glass of the booth—waving his arms emphatically at the rhythm section, as Allen concentrated on his solo choruses. (Creative a&r work, it's called.)

"You know," offered Potter at the end, "that *Biffly Blues* is the kind of piece that could hit."

"It is," said a visitor. "Anyway, it sounds just as fresh as when he first did it 30 years ago."

"No, fresher," said another onlooker softly. "Because Red is fresher. You can't date that kind of talent. And he's himself, and that means he's got things nobody else could pick up on." *(1962)*

Whir-r-r-r

At 4:00 p.m. on a quiet Friday afternoon this fall, a group gathered at the Olmstead Recording Studio to record Jimmy Giuffre's new trio. Besides the leader and his clarinet, there was pianist Paul Bley, a decidedly inner-directed man whose heavy sandals clacked noisily on the steel ladder connecting the studio and the control booth. ("Olmstead, get *that* sound on the record!" a visitor suggested.) And there was bassist Steve Swallow, barely dry behind the ears professionally—but it is said those young ears are already among the fastest in jazz.

Fast ears are what a player needs in Giuffre's music, for it is often almost totally improvised, except for a memorized statement of the theme. The players ad lib their melodies, sometimes without reference to chorus lengths, chord patterns, or any other pre-set structures. The soloist is free to shift his tempo or his key as he wishes, and the others must follow him. They do, immediately. In fact, they seem to anticipate each other even in the most unexpected turns.

The Giuffre trio had just ended a week at Trudi Heller's Downtown Versailles club in New York. Giuffre felt that after such nightly experience at improvising they were ready to record, and he asked Verve's artist and repertory man Creed Taylor to set up a session.

"It's like an instrument, this room," said Giuffre, warming up his clarinet. He had chosen Olmstead's studio himself; Verve, like several other companies, does not maintain its own recording facilities but leases them for individual sessions. Giuffre had asked Taylor for this particular one, which tops a ten-story building on Fortieth Street in midtown Manhattan. A high-ceilinged room painted in pale blue, it displays several irrelevantly ornate Ionic columns on one wall. "It's half of the old penthouse living room," says proprietor Dick Olmstead. "It belonged to William Randolph Hearst."

Olmstead's is also probably the world's only split-level recording studio; the glass-enclosed control booth is set about 20 feet above the studio floor, at one end of the room. It houses the engineer's panel, and will accommodate several visitors.

Olmstead wandered around the studio arranging his microphones. Giuffre cocked his ears for an echo after one of his clarinet phrases. "Yes, this room *is* an instrument. Listen." Swallow plucked a note on his bass and attended the reverberations.

Olmstead climbed up to the booth and slid over behind his complex board. Creed Taylor was at his elbow. Below, on the studio floor, Olmstead had placed one mike close in on Bley's piano, while Giuffre and Swallow were sharing another.

Taylor was there to represent the record company and help Giuffre. He was discreetly quiet, sensing that it was best to let Giuffre run the date himself. Unofficially present were Giuffre's wife of only a few weeks, Juanita; young Perry Robinson, a former clarinet pupil of Giuffre's; a photographer, who at first respectfully declined to enter the studio while the tape was rolling; and three friends of the Giuffres who dropped in.

"The name of this piece is *Whirr,*" said Giuffre. "Spell that," Olmstead said over the speaker, as he bent over his log.

Giuffre pretended to be puzzled. "Well, I will." He looked at the ceiling. In the booth his wife began to laugh. "W-h-i-r-r-r-r-r . . . ," he trailed off.

"Blues in B flat," cracked Swallow, naming the most basic jazz form.

They began fast. Giuffre was playing high and fingering rapidly. "He'll show those critics!" someone whispered to Mrs. Giuffre. (Because in his early days Giuffre seldom played high notes, a French critic had quipped that Giuffre's pupils needed a second instructor for the upper register.) At one point Bley hit the bass strings inside the piano with the heel of his hand to get an abrupt sound. And Giuffre got a brief effect like rushing wind by blowing across one of the stops of his clarinet.

"I think that's it," said Giuffre at the end of the piece.

Giuffre's music has obviously come a long way in the past few years. Most people who know him probably first heard of him when he wrote the highly successful piece *Four Brothers* for the 1949 Woody Herman band. This was the "cool" Herman Herd, and the brothers were a shifting foursome of saxophonists: combinations of Stan Getz, Herbie Stewart, Zoot Sims, Al Cohn, Serge Chaloff, and Giuffre. A second wave of fame came in 1957, when he formed the original Giuffre "3" of *The Train and the River, Swamp People,* and similar impressionistic pieces. Although his music was then gaining him considerable prestige and respect (he had begun winning fan magazine polls on clarinet), Giuffre gradually became dissatisfied with the restrictions his style placed on the players.

"I like the pastoral—the country," he has said of these times. "I like peaceful moods." But about then he began listening to Thelonious Monk. "I heard an element in his music—a way of stating things with conviction that was clear and sure. And he played without any restraint—he played it immediately, right in front of you. I also noticed it in Sonny Rollins's music, this same kind of statement. I got interested in this and started to work on it . . . I discovered a lot of things . . . I was holding back a lot of things . . . I was afraid of hitting certain notes . . . I worked—and finally got up enough nerve to throw the rock off the cliff and just play anything I wanted to play when I wanted to play it. It was a revelation."

In all, Giuffre spent more than a year pursuing the idioms of Thelonious Monk and Sonny Rollins. The result was the new Giuffre style that his current trio plays, but to anyone who has followed his career it also seemed like a revelation. This music incorporates the impressionistic moodiness, the textures, and the compositional refinements of the earlier Giuffre "3." But it also has a new freedom and immediacy. Now, Giuffre the studied composer and Giuffre the player who wants to make strong, spontaneous and uncompromising statements are closer together than ever.

Down in the studio, the trio was ready for the next take. "This is called *Afternoon*. It feels like a lazy afternoon," Giuffre explained, half to suggest a mood to the players. "I'll spell it for you," he said, smiling up toward Olmstead and Taylor, "A-f-t . . ."

"All right!" Olmstead smiled.

They began with no setting of tempo. Bley was soon answering Giuffre's improvised phrases in a spontaneous, far-out counterpoint. Swallow looked as if he might drop his bass, he was so acutely involved in the music.

After the number, the trio climbed up to the recording booth for the playback of *Afternoon*. Giuffre squinted as he listened.

"Let's try that one again," said Giuffre, as Olmstead shut off the tape. "For me—you guys were fine."

Bley protested, "I thought you sounded good." Giuffre quickly affected a comic pompousness: "I always sound *good,* but I can do better."

They clambered down to the studio again. Bley clacked down the steel stairs as Giuffre, behind him, suggested, "Let's leave the chord progression out of this one on the solos. I like the piece but I get tired of that progression. Let's just play."

But Giuffre was persuaded that he had played well, and they decided to return to *Afternoon* later if they felt like it.

They were now about to record *Flight.* And again they began without a tempo signal; Giuffre merely quietly said, "Okay, downbeat," as the tape began to roll.

There was a goof in the brief written introduction. "How much after you do I come in?" Bley asked.

"Two and a half beats," said Giuffre. They tried again, twice. Bley interrupted himself on the second. "If I count it, I seem to come in earlier."

"Oh, I forgot. You come in a beat and a half *before* me."

Above, Mrs. Giuffre looked at her husband and laughed quietly; she was around when the pieces were first written.

They started to play. Bley suddenly went into a medium tempo, an easy, rocking jazz groove, and Swallow, too, was on it immediately. Toward the end, Bley surreptitiously reached inside the piano again. The finish of the piece was a long note which Giuffre held softly over the piano's sustained reverberations.

During the playback, Giuffre asked Olmstead, "Dick, can we hear that piece again, on monaural?"

"I don't know why you want it in stereo at all," someone remarked. "I mean with the unity you guys have."

"It still sounds better in stereo to me," said Bley.

"Sure, *you* have a whole mike to yourself," said a visiting musician friend.

The second playback began. Swallow softly clapped his hands to the music and looked at Bley, smiling. At the end of the take, Giuffre's voice was heard, "I'll spell it for you, W-h-i-r-r-r. . . ."

Some of the pieces seemed to have little of the quality of jazz in the conventional sense. The trio's work does suggest contemporary chamber music, but only able jazz musicians deeply committed to spontaneity could improvise this way. And only musicians committed to an individual exploration of their instruments and of the personal sounds they can make with them could play this way. Classicists wouldn't qualify.

"Everything seems to be going so well," Swallow was telling Juanita, "I can't believe it. It must really be going badly. I feel like when I was in school, if I finished a test and handed it in before anybody else."

Over in a corner Giuffre was earnestly chatting with a new arrival, "At first I didn't know whether we could get along with piano in place of guitar, but Paul is fine."

"Are you kidding? You got this Swallow playing guitar parts on his bass!"

"Hey Paul," said Giuffre, turning to the door, "did you hear a string that's out of tune on the piano there toward the end?"

"On *that* piece?" Bley laughed. "Who could know? No, it was muted. I had my hand on it."

"Well," said Giuffre, as they went downstairs to the studio, "the next thing I'd like to do is record with this group and an orchestra." Bley nodded enthusiastically.

Swallow's voice was picked up from below on the open mikes. "So few notes on that last one. Sometimes with this group I sound like something from the Twenties." They ran through the theme of a piece called *Ictus* written by Bley's wife, Carla. The trio had tried to include it on a previous LP but gave up after 18 unsuccessful attempts to get a good version on tape.

"This time it's going on in one take," someone whispered to Creed Taylor. He was right.

Swallow watched intently as Giuffre invented his coda, so he would know when to come back in for a last bass phrase. Giuffre unexpectedly pulled his last note out of the air at the end of an abruptly asymmetrical phrase, and Swallow was right on it. Bley concluded with something of his own.

No sooner had Olmstead shut off the tape than everybody broke into laughter at the humorous appropriateness of it.

The playback got immediate approval. "Now I think this next one should be very loose," Giuffre said, throwing his arms in all directions. "You know—very nondirectional. This is called *The Gamut,*" he announced to Olmstead.

During his solo, Bley began to hum his melody as he ad libbed. Neither Olmstead nor Taylor objected. They both agree with Giuffre that it's part of the music as the trio makes it.

Again, Giuffre listened to the playback with his eyes half shut. This time Swallow's eyes were on him, as they had been while they were playing. Suddenly everyone realized that all three of them at one point had spontaneously fallen into playing a little traditional jazz phrase. "Yeah, that's a good figure there," said Bley quietly. Later he turned to Swallow. "You see, when we set him up, that Giuffre really plays a good solo." They laughed and then listened carefully to the way Bley had finished the piece with an unusual sound from the piano.

"We ought to end with that," said Giuffre.

Creed: "You mean end the *record?*"

"Yes," said Giuffre and Swallow in unison.

Giuffre looked at his watch. "Hey, we've still got the studio for half an hour, haven't we, Dick? Let's hear it all back again. We're all finished. We'll call this album *Thesis.*"

"Finished *now?*"

"Sure. After all, we only did one piece twice; there isn't much choice to make, and there's no editing. I'll spell it for you. F-i-n-i-s-h-e-d." (*1962*)

Stitt in the Studio

Alto saxophonist Sonny Stitt was one of the first jazzmen to grasp Charlie Parker's style, and he had apparently recognized an aesthetic kinship with Parker long before most followers of jazz had heard of either man. His absorption of Parker's ideas is so complete that, as one commentator has put it, Stitt is the kind of player who refutes every concept we have about originality, even personal expression, in jazz. Yet Stitt plays with spontaneity, involvement, and conviction. If he lacks Parker's brilliance and his daring quickness of imagination in rhythm, harmony, and melody, Stitt nevertheless is not playing an imitation, and his work is far from pastiche or popularization. He simply finds his own voice in Parker's musical language. He may construct a solo almost entirely out of Parker's ideas, but he will play them so as to convince you that he discovered each of them for himself. As if to give a final contradiction, Stitt also plays tenor, usually in different, somewhat simpler style, but with no less effectiveness. By almost all we profess to believe about jazz, it cannot be. But it is.

Stitt has recorded in several settings, but most often with himself, piano, bass, and drums. He has most often appeared that way too, sometimes touring alone and picking up local rhythm sections from place to place. He recently undertook to record for Atlantic, a new company for him, and everyone agreed that it was time to try Sonny Stitt in a new setting. Thereby the date obviously held promise. It also involved some compromises. Stitt was to be provided with a large group consisting of ample brass and rhythm sections, and with scores by Jimmy Mundy, who was doing classic pieces and arrangements in the early thirties for Earl Hines and then Benny Goodman, and by Tadd Dameron, one of the first and best to do big band modern jazz arrangements. But some weeks before the date, Stitt had made some appearances with the modishly successful sax and electric organ setup. So the rhythm section was to include organ instead of piano. Also, Mundy's scores proved to be capable of reworkings of blues phrases and ideas—probably what the job called for.

As the musicians gathered at Atlantic's studios, it was obvious that the chairs assigned to the three trumpets, one French horn, three trombones, bass, drums, and organ would be filled by some well-known players. Not the least of them was Philly Joe Jones, currently one of the most celebrated drummers in jazz.

Sonny Stitt entered the studio, a tall and almost unbelievably thin presence, graying slightly at the temples now but otherwise looking still like a man in his early thirties. He spied veteran trumpeter Dick Vance first and, more given to gestures than talk, immediately fell on Vance's neck.

Soon the altoist was warming up, not with scales or exercises but with some of his own favorite phrases, and Mundy took his position in the center of a semicircle of horns. Nearly thirty years of jazz music were presented and ready to go.

Mundy did not seem to be carrying himself exactly with the assurance of a veteran among these mostly younger men, but he was obviously impatient to get to work. "Let's go. Let's try *Boom Boom*," he said, calling for one of the scores which had been placed on music racks before the players. And turning to Sonny, "Play me a D minor chord?" Stitt obliged with an arpeggio. "Ok, everybody ready?" Mundy asked, raising both arms wide apart, and to about three-quarter mast. "I'll give you four bars for nothing," he said, beginning to count off, and the players came in for a very ragged start. When they began it again, the piece proved to be assertive and based on a simple blues bass figure.

"Now let's try it just a trifle brighter," said Mundy, meaning in jazz language that he wanted it a little faster.

Stitt briefly crossed to the corner of the studio to bum a cigarette from a visiting friend, remarking with the hint of a smile, "He's got some screaming brass there, hasn't he?"

Mundy was addressing Atlantic recording engineer Phil Iehle, "Can we get the French horn in closer, please? It's isolated over there away from the other brass."

As Iehle moved the horn chair, Stitt, now back in position, made his request, "I don't dig this sitting down playing."

"Oh, you want to stand up?" Iehle raised Stitt's microphone.

The reshuffling had an obvious and immediate effect on the next run-through of *Boom Boom;* the written parts went down with a spirited crispness and Stitt was really beginning to play the blues. Mundy looked up. "Want to take one?"

"Yeah," Sonny agreed, "let's take one."

And Phil Iehle began to roll his tapes and receive the first try at *Boom Boom.*

At jazz record dates, the end of the first take is often an unofficial signal for a brief break. And this break was marked by the arrival of Ahmet Ertegun of Atlantic, who was to supervise the session. Ertegun is the son of a Turkish diplomat and, along with his brother Nesuhi Ertegun, he has been involved with jazz most of his life. He entered, announcing to the assemblage with a smile that, of course, it is musicians who are supposed to be late for record dates, never producers. He shook hands with Mundy and Stitt and joked briefly with Jones. Then he crossed the room to greet Tadd Dameron, who was sitting with a couple of friends, waiting his turn after Mundy's.

Ertegun heard the first playback in the engineering booth, instructing Iehle, "Phil, this is basically Sonny Stitt, rhythm and organ. Let the brass be more like punctuation and bring up the organ especially." In the studio, on the other side of the glass panel that separated the players from Iehle's tape machines and control board, Mundy stood with his ear glued to a loudspeaker. The rest of the room, including Stitt, strolled around and chatted, apparently unconcerned with what the playback was revealing. But they were hearing it nevertheless.

When the speakers were silent again. Ertegun, Iehle, Mundy, and Stitt had heard how things were going, and the date was on in earnest.

Another take. The organist, a capable young girl named Peri Lee, who knows all the hip phrases, was playing harder now, but the brass was so pungent as to balance out. And from time to time Stitt even improvised on top of some of the high brass screams, gracefully unintimidated.

At the end, Iehle and Ertegun wanted to test the microphone balance of drums and bass, and Philly Joe and bassist Joe Benjamin began to play alone as Iehle moved his mikes and adjusted his settings. But almost immediately Stitt got interested and was playing the blues along with them.

Minutes later, they had done an almost perfect take, and Ertegun said quickly over the studio speakers, "Can you do another one right away?" They did. Stitt stood up for his solo, played with his knees slightly bent, rose to his toes for each of his high notes, leaned back as he descended, and was on his toes again as he reached for his ending.

Ertegun: "OK, I think we got it."

Mundy's final blues had effects as fashionably hip as its

title, *Soulville*. "Ahmet," he said into a studio mike, "can we keep the organ up? And can you fade out the ending in there on the board?" They did a run-through and a take, Mundy exuberantly conducting with both arms. Mutually, Mundy and Ertegun suggested the piece could carry more improvising by Stitt and a solo by trumpeter Blue Mitchell as well. As the tapes were rolling again, Mundy hurriedly expanded his arrangement, making signals to the men to return to section "B" by holding up several fingers on both hands to describe an awkward "B," or to repeat "C" by inscribing a "C" in the air with his right fore-finger.

At the end, Ertegun from the booth made a long announcement over his microphone to the effect that it sounded like the rhythm was slowing down at certain points, although it might not actually be, etc. Philly Joe Jones, apparently having heard all he intended to listen to, started talking loudly to the speaker on his right, "OK, OK, OK, baby! OK, Ahmet! OK, Ahmet (this time pronouncing it *Ak*-med)! When we stop the shuffle beat and go into the four it sounds like the rhythm drops, see? But it doesn't. See? OK, OK."

In the next take Stitt was playing strong blues, and he was carried away enough to improvise an extra chorus. Mitchell followed him and everyone else made necessary adjustments, with Mundy's arms flagging the assemblage on, 1-2-3-4, 1-2-3-4. And Jones fixed things so one didn't notice the change of rhythm, although he still made it.

At the end Ertegun entered the studio. "All right," Stitt said, crossing over to him, "you like that one?"

"Yeah!" They shook hands.

"That's it, then," and Stitt returned to his position in the semicircle of musicians, pausing en route long enough to give Mundy a polite embrace of thanks.

Meanwhile, Tadd Dameron, a keen, sharp-eyed man, had

been distributing his own music around the room. His first piece had been given the hasty, last-minute title of *The 490* after Dameron's street address. It was, by Ertegun's request, an eight-bar blues. But the musicians' first run-through revealed that Dameron had written an ingeniously ballad-like theme within the traditional but often neglected blues form, and had scored it quite interestingly. Once he had signaled the tempo and got the musicians started, Dameron sat quietly facing them, listening to the results with sober face and piercing look. Only once did he raise his hands, to quiet the brass behind Stitt's first entrance. And the room had taken on a new life.

At the end, Dameron was immediately exhorting the trumpeters to, "Sing it! Sing it! Everybody!" He started them again, this time knowing what to listen for and conducting them modestly with one hand. "Hold it! Hold it!" he stopped them, turning to the trumpeters. "Play it this way, boo-ob-de-*wahhh*."

"Hold it, Sonny," Dameron smiled an interruption toward the end of another run-through. "That's the intro."

"Oh," said Stitt. He had momentarily been improvising with his eyes shut; as the music had become more challenging, he had become more serious. "I was back at the second ending?"

"You were wailing through, man," said Dameron, signaling a general laughter.

They tried *The 490* onto tape. Dameron smiled at Philly Joe's exciting and propelling delayed entrance. Gradually, the composer had begun to do more active conducting. But he was not keeping time for them so much as he was signaling dynamics and encouraging feeling, and he frequently sang along silently with the brass as Stitt effectively juxtaposed passionate slow blues improvising against the more sedate ballad-blues writing. In a peak in the coda, Sonny lifted his left foot abruptly behind him, and Dameron held his finger to his lips for the last big chord from the brass.

"Well!" announced Philly Joe at the end of the playback, "that ain't no easy tempo to keep going."

Dameron began to work with the players on the next piece, *On a Misty Night*. Again he was calm and strictly business, and again the musicians were quietly enthusiastic. "More legato. Make it slurred more right there. Good. Thank you," he said interrupting them and singing the passage for them twice.

Stitt's part, of course, was mostly inscribed as a sequence of chord symbols on the music in front of him, and he was to improvise his melodies to fit them. But almost from the beginning, Stitt was simply playing without referring to his written part: he didn't need to; he had glanced at his sheet and heard the group play the piece through, so he knew what the chords were almost automatically. Except for a tricky introduction or ending, he just played.

Dameron increased the tempo each time they ran the piece through, and a huge, but decidedly lyric, brass sound gradually emerged. For one crescendo he raised himself to his full height, moved forward and conducted the semicircle with both arms. Stitt had begun to sit down during these rehearsals, but he was playing like Sonny Stitt standing up nevertheless.

"Trumpets, it's supposed to go do-*wah*. You're playing do-*dah*. Dig?" Stitt's coda was of a length dictated by his own inspiration, and Dameron had to bring the brass in under him for the ending when he intuited that Stitt was ready.

"How long is that tag?"

"Till you get tired," Dameron answered quietly, laying aside one of his several white handkerchiefs.

"What's my chord?"

"B flat 7."

Each time Stitt made a different ending and each time Dameron knew when to signal the group to re-enter. But, when the take was under way, Stitt made a mistake at the end. The rest of the performance had been so good they decided to do just the ending again and splice it on.

"That's it," Ertegun announced through the speakers. "But Sonny, Blue and the rhythm section, can you stick around a second?" The idea was to do one informal quintet number.

Two minutes later, Stitt had taught Mitchell a little thirty-two-bar theme he had apparently made up on the spot by expanding a little indigenous phrase, and the rhythm section had fallen in with them. Six minutes after that, the quintet had played it and improvised on it expertly, and Iehle had the results on tape. Then they tried it a second time, but it didn't come off. So, by unspoken agreement, they decided to accept the first version as much the better.

Everyone knew then that the date was over. The musicians joked a little as they packed up their instruments, and Iehle silently disconnected his microphones. Then everybody left. (*1962*)

Rehearsal Diary

The Tuesday afternoon gathering is at Carroll Instrument Studios on West Forty-ninth Street. The occasion is the first of three rehearsals for the final concert in the Carnegie Recital Hall series "Twentieth Century Innovations" directed by Gunther Schuller. The series began with Stravinsky, Schönberg, and Webern; it moved in later concerts through the American premiere of several concert works (including one by Barraqué); and it is to end with "Recent Developments in Jazz." Originally, some "third stream" pieces were to be presented, but these were combined with the "Early Experiments in Jazz" of Stravinsky et al., at the previous concert.

Schuller stands in the brightly lit room, intently facing the players, his scores spread on a music stand before him. Some well-known jazzmen are present. Nick Travis and Don Ellis are on trumpets, Britt Woodman and Jimmy Knepper on trombones. Among the reeds are Phil Woods and Benny Golson. Drums and percussion are Charlie Persip and Sticks Evans. There are two basses, Richard Davis and Barre Phillips. Jim Hall is on guitar.

"Let's go," Schuller is saying. "Get out the Lalo Schifrin piece," calling for the toughest work they are to do, a twelve-tone serial composition, *The Ritual of Sound,* involving jazz motives and phrasing. They discover there are no markings of sections or subdivisions on the score. "You'd think a big copying house would question a thing like that," Schuller observes.

"Well," (it is a voice among the brass) "there are so many metronome markings—why don't we just use those." The shifting rhythmic complexities of the piece call for indications scattered throughout the score.

"At the first two-four, let's call that A. B eleven bars later. C at the first tempo change . . ." The players mark their scores as Schuller continues. "Wait a minute. I'm calling this off, but not marking it myself."

"Where was B, Gunther?"

"Lalo's here! He just walked in." He is a quiet, sober presence of medium height, dark, and currently slightly stout around the middle.

The piece, like most serial works, has notes and melodic fragments scattered from instrument to instrument in waves that carry us across the orchestra and back. But much jazz phrasing is called for, and not only in the few obvious growls and shakes from the brass.

"Right there before D, it's *do-wat, do-wat.* You know, like the Ellington thing," Schuller instructs. Then to Lalo, "This is pretty fast." (To the left of the room, a good-looking female enters quietly.) "Is the trumpet open there, Don?" he says, stopping Ellis.

"Oh, you're right. Still cup-muted," says Ellis as Schifrin peers over his shoulder at his part. Down in front, among the reeds, Benny Golson looks at his part, bemused but interested. Over on the right, Richard Davis is smiling—but then Richard Davis is frequently smiling.

"Now listen, all of this section is jazz, remember. Play it *do-la-DAH!* But basses, you're rushing. Try that again, just the basses. No, watch the markings. It's very slow here."

Then, a voice from the center of the group, "Gunther could we go back to G?"

"No, I don't want to rehearse it now. I just want to read it."

They are at about the middle. The tempo changes radically, yet the musical momentum seems continuous. "Hold it! That's too slow, but, Richard, you got there too early."

And a few moments later comes a topper: "Everybody, at L it gets complicated. But Lalo, aren't there some wrong notes in here?"

Schifrin comes forward and stands beside Gunther, addressing them in his slightly Spanish-accented speech. "You see, any one of these phrases would be an *ostinato* riff in jazz or in any African-influenced music. But I have gone from one to the next instead of repeating, and given each one its own rhythm, separately."

They are into it again. The reading is still rough, but in the places where it should, it is beginning to swing a bit.

"Hold it, hold it. Come on, you can read better than that. You'll just have to count."

And finally, when the last notes have hesitantly scampered by: "We'll have to spend a lot of time on that tomorrow. Now, get out George Russell's *Lydian M-1.*"

There is some shifting around: Persip moves into the drums; the reeds move over to include Phil Woods; Schifrin takes his place at the piano; Travis is out; etc. They begin. The intro is Persip setting a quick tempo. Then Russell's theme, a curiously compelling rapid succession of notes, tightly knit, and giving the impression of a narrow range melodically, but a wide one rhythmically. It must be tough enough to read them, tougher still to get the phrasing and accents. "Hold it! You can't read this fast?"

"No. Gunther, you'd better get Barry."

"No, look, Barry isn't going to be here. Can't you . . ."

"Okay, I'll work on it tonight and have it tomorrow."

"So far, so bad," mutters Nick Travis on the sideline.

Soon Gunther is saying, "Let's just take a break and then we'll do the Hodeir pieces. I want to read through everything today." Only about two-thirds of the players leave for the Coke machine in the next room. The rest are running through tough passages they have already played, or running through the opening section of André Hodeir's *Jazz Cantata.*

Out by the hall phone booths someone is saying, "You know, the trouble is that if you sent down to the corner music store for the worst pop song in the house, handed it to Thelonious Monk or Sonny Rollins and said, 'Want to try this?' in ten minutes he might come up with something that would knock you out." (He has a point, and a big one, but one that every good jazz composer is aware of.)

Inside, Gunther gathers the players, calling out for the *Cantata.* "And remember the phrasing—most of this is just plain old jazz." A good-looking woman takes her place beside him. She is of medium height, with almost black hair and fair, almost white, skin. She takes an authoritative stance as she holds up a vocal score at eye level.

The first section is very fast. She lets go with a darting, wordless, scat melodic line, very high for jazz singing. Behind her Phil Woods has a rapid alto part to play in the style of an improvised solo. The singer is dashing off her notes with a combination of ease and drive.

At the end of the first section, she protests, "I didn't get all the notes." She is saying it over a spontaneous scattered applause from the musicians.

"You stayed with it all the way through!" Gunther counters. "Gentlemen, this is Susan Belink. She has never sung jazz before. She's in opera."

They are into the second part of Hodeir's *Cantata.* The performance moves along well enough, but at one break Jimmy Knepper says, in a near-whisper, "That note is awfully low." Several of Hodeir's parts skip rapidly from high to low and back again.

Gunther asks, "Have you got one of those long fiber mutes? Use that." Britt Woodman, to Knepper's left, silently provides one.

Soon they begin Duke Ellington's *Reminiscing in Tempo,* first done in 1935 and one of his early long works. At one brief interruption, Schuller ejects, "Hey, this sounds good!" And at the end, "Good. I think we can get that one. But somewhere along the line we're going to have to add some more rehearsal time."

"How about starting at one o'clock Thursday afternoon instead of two?" someone suggests, to general agreement.

"Right now as much as we can of the other Hodeir pieces, *Paradox I* and *Tension-Détente.* From now on I'll just call that *Tension,"* he says, pronouncing it in English. "Anyone who doesn't have a part in these can leave."

"Trouble is," one of the players is remarking to a friend

as he heaves on his overcoat, "that in jazz the players always come first. No matter how well we do that Ellington piece, it was still written with Harry Carney's baritone sound in mind. And Johnny Hodges' sound, and Cootie Williams' sound, and Rex Stewart's sound. Don and Nick are doing especially well with those trumpet parts, but no matter how well any of us do . . . well, it's a challenge."

"Still," says the friend as they push through the door, "what about George Russell's piece? He didn't necessarily write that for specific men. True, it was commissioned for that Teddy Charles record, and he more or less knew who would be playing it. But he still wrote it sort of in the abstract. Any group of good jazz players ought to be able to play it as well as any other group."

"Yeah—theoretically . . ." They are both smiling as they reach the street.

Wednesday, 9:00 a.m. Second rehearsal. At Carnegie Recital Hall itself. The small auditorium seems hazily lighted at this unaccustomed morning hour, but the stage is harshly bright. Almost everyone is prompt and assembled, instruments unpacked, on stage by five minutes to nine. And at five minutes past, Gunther is saying, "Oh, Jim, there you are. Fast as you can."

They begin Schifrin's piece again. Gunther conducts with both arms, but somehow always with his right more than his left. He bobs up and down, bending his knees a bit, almost awkwardly, and moving slightly from side to side. "Remember, all of this has more jazz phrasing than we gave it last time." Hall is crawling around the stage as they play, searching for an electrical outlet for his guitar amplifier and trying to trail his wires so that no one will trip over them.

Schuller stops everyone, looks down at his score, lays down his baton. "Start that again, but just the trumpets."

"Okay. Now one trumpet at a time. You first, Nick."

"Now just the trumpets and the trombones."

"Okay, good. Now the whole six-four section, all the brass." Breaking it down had worked.

Travis signals a private joke to Phil Woods, who is sitting in the auditorium, and they both laugh broadly but silently.

At the end of a wildly contrapuntal texture, there are pleased looks all around. Schifrin, who had been standing in back of the house, makes only one quiet comment to Schuller, so quietly that I don't catch it.

"Now, just the guitar and vibes on the eleven-four." Sticks Evans is having trouble with this part, but even so he almost makes it swing. Too infectiously perhaps, because by the end of one section Schuller is saying, "Now fellas, here don't play swinging eighth notes. These are all legit eighth notes."

Out in the auditorium there is some reasonably good-natured joking to the effect that, "Maybe Lalo sniffs air-plane glue or something before he writes something like this."

10:00 a.m. They are still on Schifrin's piece. "Did I tell you to shorten those notes yesterday? Well today I'm not so sure."

"Britt, those triplet accents . . ." Gunther sings the phrases, and Travis spontaneously begins to play them on his trumpet.

Jim Hall, asked for a particular dotted effect, says, "It's been so long since I played a mandolin."

By 11:05 the final notes rise in an orderly pattern and hover over the stage, the composer applauds them from the back of the house.

"Okay, in five minutes we'll do the Ellington." He says it to the men in the auditorium as well as those on stage, and some of the former begin to take their places with the group.

Phil Woods warms up by playing *Donna Lee*. Gunther is telling Britt Woodman, "I'll rely on you here."

"But I never played this piece with Duke."

"But you know the style."

"Where did you get the score?"

"I asked Ellington for it a few years ago to use in a concert at Brandeis University. He never sent it. Maybe he couldn't find it any more. So I just took it off the record."

They are into *Reminiscing in Tempo,* playing it with only a couple of interruptions and corrections. "Soft," says Gunther over the music. "Full!" "Relax." "Up!" It is as if Ellington's gentle but firm melody had a momentum of its own that carried the players along with it. "Hey that sounded great. We'll do this some more tomorrow."

11:20. "Real quick, get out the George Russell. Wait, somebody's missing." He faces the back of the house. "Hey, Buff, is Harvey out there?"

After two false starts, they have slowed down Russell's rapid tempo to work on the accents. "Lalo, you'll just have to play louder. I can't hear you."

"Charlie, you're rushing. No, don't look at me in this piece. There's no point in my conducting a piece like this anyway. Just watch for that one bar in three-four. Otherwise, keep playing all the way through the tempo I give you in the beginning."

At the end someone asks, "Could we add another rehearsal?" He sounds almost plaintive.

Thursday, the afternoon of the concert, 1:00 p.m.

"Get out the George Russell," Gunther announces loudly and hopefully to the harshly but brightly lighted hall. Not all of the musicians are there, and only half of those that are seem ready. Gunther crosses for a consultation with Jim Hall over the score.

In a few minutes, they're assembled. Schuller signals the tempo. Persip's brushes chip a snare drum. "Wait, wait. Hold it. Charlie, you started faster than I set it." ("Oh? Sorry.") Persip cooks up his brushes again. Everything is falling into place. Hall is laying down his guitar line with confidence. The difficult horn accents are coming more correctly.

"Let's stop there. Charlie, help us out there with some fills. If you play right through, you leave a hole. Now, begin again at M for Monk."

"Very big on those D's. At the end, Lalo and Richard louder to help everybody out. Make a big thing of it!" They do the final part again.

"Okay. Now from the top, just guitar, piano, vibes, bass, and drums. No horns. We've got to get this. First we'll try it very slowly." It sounds good slow.

"Hey! This is a hell of a piece!" says someone suddenly as they finish the first marking.

Schuller, emphatically: "There is not one extraneous note in this whole piece."

Golson, laughing: "Except in the solos!" He had just played one.

Then, George Russell's *Lydian M-1* from the top. It goes down very well, the accents fall right, the players are more relaxed, and a group drive has developed. Schuller seems cheerful for the first time today as he calls out for the Ellington piece.

"Play it more evenly, Lalo. And these ensembles have got to be right. It's like playing Mozart—everybody knows the style." Again, a very good reading.

But Gunther is a bit strained as they turn to the tough one, the Schifrin *Ritual of Sound*. "Please concentrate. I have no more time, and I've got to do the Hodeir pieces. Where's Phil? Benny, is Phil out there?"

"Joe, always strong!" They are about halfway through it now. "Always with energy." Schifrin himself is in the rear of the house quietly commenting to a friend in Spanish—a good sign presumably.

"Drums, you're late. *Wum-b-um-bum*. We'll work this

out later between us." Travis makes a mistake—his first, as I remember.

At a final run-through, it goes the best yet.

"The Hodeir. The Hodeir now," clapping his hands and beginning to shift a couple of chairs around. As they begin the first section, Susan Belink joins them, singing her part from a seat in the house. Schuller asks her to hold it, he wants to rehearse the players alone first. "Don, watch your dynamics."

Then Miss Belink moves toward the stage. She is dressed quite informally compared to her first appearance. And today her sentences feature a couple of scattered "man's" and "like's."

In the first section she and Woods make the shifts of tempo together in sudden, compelling bursts. Her head tosses back slightly on high notes, and Woods's feet lift and curl up under his chair as he plays. At her last strong note, a sudden burst of "Wow!" goes up from the group, and the whole room is smiling broadly.

It is nearly three o'clock. An engineer has set up a mike which somehow picks up more Woods than Belink, although she is singing directly into it. They move singer and mike to the center of the stage. Gunther suggests to Don Ellis that for his trumpet solo in the fast section, "We have the mike, so use the Miles Davis mute."

"Yes, well, I'll try the Harmon, with the stem in, close to the mike." Somehow it doesn't work; it doesn't get the Miles Davis sound. "At home I have a Miles Davis bucket mute and a Don Goldie mute and a . . . Let's try the cup mute." Gunther goes out into the auditorium to hear the balance again now that the mike is on Ellis.

They begin the first section of the *Cantata* again. As Belink brings out her last phrase Gunther turns to the visitors to ask, "Could you hear that?"

"We couldn't miss it!" "Sure did!" (Visitors now include

composer Milton Babbitt and tenor saxophonist Lucky Thompson.) There is a five-minute break, and Golson immediately climbs off the stage and goes over to Thompson, seated in the rear of the house, with a warm welcome. Several players exchange tips on the food at nearby lunch rooms as they hurry out.

In about five minutes the players are reassembled on stage for the *Cantata* again. It goes excitingly and there are smiles and scattered applause at the end.

4:20. "Okay, the other Hodeir pieces. Get out *Tension*. We're really going to be hopping it. And we're not going to rehearse past five. The rest of you, who don't play these last two pieces, be here tonight by 8:15 at the latest. And have a good dinner. It helps. Wait! Before you go, just let me run down the final number. You know, Eric Dolphy and his quartet will be on in the second half of the program, playing their own things. At the end we want to have an improvisation for almost everybody on *Donna Lee*. Solos by you, Phil, by Don, Benny, Jimmy, Nick. But don't anybody play the theme in the first chorus unless you really know it. When we finish the solos we play the theme again. But don't take it out, because then Eric will play two choruses very high, you know the way he does. Phil, you improvise at the same time, but in the middle. And you guys," gesturing to the other horn men, "play a riff under them. Okay? Now let's go on to *Tension*."

In the back of the hall Milton Babbitt whispers, "So *that's* how they arrange an improvisation!"

On the stage, Gunther is instructing, "Joe, remember it is a vibe solo, but he's written it out. Try to play it like you're improvising it."

"Richard, you know that little hole in there, give me a Ray Brown walk. And everybody, you know this little measure I'm always bitching about. Softer. Not *da-da de-du dit-do-WAH*, but *da-ut da-it dit-da-do*. Begin again."

4:40. "Okay. Now, *Bicinium*. Don't goof now." He is speaking quickly and almost sternly, but suddenly he laughs, turning to me, "Martin, are you getting all this?" It breaks the tension.

Again, Golson and Knepper have to move quickly from high notes to low. Their chops must be aching. Persip claims he's getting blisters on his feet. "Come on," Gunther encourages. "Just imagine you're back with Dizzy. He worked you much harder than this."

4:55. "Let's take a breath and we'll do it again. We have another one to rehearse after this. Good music is always hard."

"You *mean* that?" says Travis, genuinely puzzled. "I've played good pieces that were easy."

"When?"

A pause. "Oh, I remember one in about 1948."

"Just these three rehearsals and the cost is $2,500," Schuller muses. "Sometimes I think all regulations are deliberately anti-art."

And in a few minutes, "Okay, now *Paradox I*. Only Charlie, Richard, Jimmy, and Benny. The rest of you can go. By the way, this will be the first number on the program." (*1962*)

Record Date: Art Farmer and Jim Hall

Trumpeter—and nowadays more frequently fluegelhornist—Art Farmer has formed a quartet featuring guitarist Jim Hall, and the alliance promises to be fruitful. Both players are lyricists. Farmer, as a direct heir to the innovators of the late 1940s, has his own kind of virtuosity on occasion, while Hall, his own roots more obviously stretching back to the late thirties, is a somewhat gentler player. Therefore, there is likely to be musical empathy, plus good contrast and little that is stylistically redundant in an alliance of Farmer and Hall.

Farmer is, in style and temperament, a modernist. He can undertake almost any musical task—reflect any passing fancy from cool through soul and almost any Tin Pan Alley ditty—with honesty and integrity, without calculation or compromise.

Another musician might tone down his style to the point of inhibition on this or that number. But Farmer simply does his straightaway best on whatever material he undertakes. It is a rare quality. And it should go without saying that Farmer's best is something special and personal.

Soon after its inception, the Farmer quartet was signed by Atlantic records, and its first LP was undertaken in late summer.

Farmer's determination that the record should show the quartet at its best, plus the use of some still-unfamiliar material on the LP—not to mention Atlantic's usual care in recording—led to a series of three recording dates. And before it was over, the third of them proved singularly fruitful.

It is an evening session, scheduled to begin at 8:00 p.m., in the Atlantic studios. By 7:45, drummer Walter Perkins and bassist Steve Swallow are on hand. Perkins has his set assembled, and Swallow already has been "miked"—a microphone, wrapped with foam-rubber padding, has been tucked into the bridge of his instrument by Atlantic's engineer, Tommy Dowd.

Hall is outside the studio door chatting with a friend

when Farmer steps off the elevator, shakes hands all around, and turns to the studio door with the apology that he needs to warm up.

Inside, Farmer's warming up soon proves to include not only exercising fluegelhorn and lip (with some George Russell scales, by the way), but also his learning a new piece by running it over attentively on the studio piano. It is a gently appealing waltz called *Some Sweet Day,* which sounds as though it might be Jim Hall's. However, Hall explains it was written by a friend, a composer-singer from Argentina, Sergio Mihanovic, whose family comes from Yugoslavia.

Bassist Swallow complains that he can't seem to get an unwanted buzz out of his E string, as Hall, his guitar and amplifier set up and properly microphoned by Dowd, consults with Farmer.

Perkins has begun energetically demonstrating a tambourine he has hopes of using on a piece by Tom McIntosh called *Great Day*. Farmer's "I don't like the sound of that—play it on the drums" is met by an almost crestfallen look from Perkins.

"Aw, I practiced all week getting my technique down," he says.

Swallow is rehearsing the first piece (not by running over a bass part of chord changes but by playing the melody itself), and it now appears that composer Arif Mardin, assistant to Nesuhi Ertegun, jazz a&r man for Atlantic, and Ertegun himself, have arrived.

Ertegun delivers his greetings as Swallow and Farmer are running down the skeletal arrangement of McIntosh's piece. About the ending, "Make a cymbal bash on that big note," Farmer instructs Perkins, who sits at his set, surrounded by three microphones and an appalling tangle of wires. Perkins gives a sample of his best bash. Then Farmer continues talking to Swallow and Hall in his usual firm understatements: "Yeah. And y'all play the big fat chord."

So the ending is set. They go back to the first chorus, but that in turn leads to further changes. ("We're going to have to change that tag." "Let's simplify and play more unison.")

Soon they are running through *Great Day* from the top. Farmer plays a lovely solo, an example of his special unruffled sprightliness. He is undoubtedly only half trying—just running through the arrangement—and he plays as he walks across the studio to resume his place at his own microphone after a consultation with Hall. But it sounds lovely.

McIntosh arrives, entering the studio with an apology that he knows his piece needs some changes, but Farmer, smiling, immediately reassures him that they had been working on it and he thinks Tom will like what they had done.

Great Day, it turns out, evokes a happy spiritual, and thankfully proves to be without the affectations and clichés of soul music.

Farmer takes a final run-through. There are a few brush-up corrections ("Walt, do something harder on that last note") before Ertegun suggests from the engineer's booth, through the loud-speaker, "Let's try it, Art."

"Okay."

"Stand by. Here we go. Seven one three four, *Great Day*. Take 1."

The performance unfolds.

Hall, without being derivative about it, suggests a contemporary Charlie Christian.

At the end there are two further changes in the arrangement, and a two-bar break goes back in, and Ertegun voices approval of the sound and balance Dowd is getting after they hear the playback.

"Jim, you have to get out your rock-and-roll guitar to play that last note," Farmer chides. And then through his microphone he addresses Ertegun and Dowd: "Ready to try another one?"

"Any time," Ertegun responds. And as Dowd nods, he

formally announces for the benefit of the tape, *"Great Day,* Take 2."

As they play, Swallow is curved around his instrument, standing on tiptoe as if he were about to climb up its side. Hall looks as unruffled as always and plays calmly but feelingly. And Farmer has taken off his loafers—undoubtedly a good sign.

At the end of the take, Ertegun encourages over the studio loud-speaker, "Very good except for the ending. A beautiful take otherwise."

But Farmer doesn't quite agree and says:

"Let's try another one on it. It's still a little tight."

After a pause he adds, "Hey, Nesuhi, could you play us back a little bit of the first chorus."

Soon they've heard the playback, commented on the arrangement, and are about set for another take.

As he leaves the studio to re-enter the engineer's booth, McIntosh pauses and says, "Let Walt have two bars after the solo." And as Farmer thinks this over, he adds, "You got two good takes already though."

"Good," says Art—and as a sly afterthought, "they can use them in the memorial album."

During the last take of *Great Day,* Swallow adds some further toeing and dancing to his bass climbing . . . and as the last notes die, Ertegun asks for an extra ending, to make up for the one that didn't do so well. That done, the studio fills with casual conversing and a bit of laughter.

Gradually the business at hand begins to re-emerge as Hall recalled the previous two sessions with a question to Farmer, "Did we ever get a good take on *My Little Suede Shoes?"*

"Yes, we finally did last time."

"Art"—it's Jim Hall again—"do you want to do the ballad?"

"Why don't I join in?" asks Swallow, glancing at a copy of *Some Sweet Day* on Hall's stand. "I can read off of Jim's part."

They begin slowly and quietly. Behind the glass of the booth, Dowd is readjusting his levels to fit the new dynamics. He operates his control board almost as though he were playing a piano, using several fingers of each hand simultaneously on the various colored buttons, adjusting them to raise this microphone level or lower that one.

"That buzz in the E string on the bass is still giving us trouble," he remarks to Mardin.

At the end, Hall says, "That sounded much too serious. It shouldn't be a driving waltz."

Another try. Perkins uses his brushes to much better effect.

Dowd, meanwhile, is responding to some kidding about having three mikes on the drum set. He says that, usually, little of the true quality of jazz drumming comes through on records. "All you have to do on some dates is go in the studio and hear what's really being played, then listen to how little of it the tape is picking up."

At which point Ertegun reminds him, "Tommy, we *have* to get that middle cymbal."

When the take is over, Ertegun says into his mike, "It could be a little faster, Art."

Farmer nods.

Another try at *Some Sweet Day,* and from the opening bars, it is obviously going to be a good one. Swallow sways widely in time to his own solo, causing Dowd to remark, "It's a good thing we've got the mike strapped inside the bass, otherwise we might not be getting but half of this."

McIntosh looks pleased at the whole performance.

In a few minutes, Farmer is in a huddle with Perkins across the studio as Ertegun calls over the loud-speaker, "Art, I think you should hear this played back!"

At the end all agree it was a beautiful take, and Farmer affirms that he is now relaxed about the way the piece is going by casually telling an anecdote about Dizzy Gillespie's fluffing a note during a performance of the *Star-Spangled Banner* at the Monterey Jazz Festival.

However, in a few moments, Ertegun suggests, "Just for safety let's do another one—and maybe we'll use the first one at that."

Hall solos with his eyes shut, embracing his instrument. And in his solo Farmer shows that on fluegelhorn he gets the intimacy of a Harmon- or felt-muted trumpet with the warm sound of an open horn.

A few minutes after the final take of *Some Sweet Day,* Farmer and Perkins have begun to kid around in duo with the theme of a light and brash Richard Rodgers ditty called *Loads of Love.* The kidding comes off so well that someone is soon suggesting they record the piece as a couple of duets—first Farmer and Perkins and then Hall and Swallow. They run it down that way a couple of times, but it seems to be falling into two separate halves. Farmer and Ertegun suggest that Perkins continue his sock cymbal under a solo by Swallow and then under Swallow and Hall in duo.

"Want to try it a little faster?" Hall asks, looking up at Farmer.

"Well, if I have to start worrying about the tempo and the changes too. . . ."

Meanwhile, Dowd has made it a real evening session: he has turned out the studio lights except for a couple of pin-lights on Perkins's drums and for the much softer glow through the glass of the engineer's booth. It is as if he were deliberately preparing for something.

A couple more takes of *Loads of Love* and Farmer is asking Perkins, "Can you keep the feeling of 'two' but play around it a little more? I think of it as a solo almost." After several false starts, they get a couple of good takes on the piece, and Ertegun suggests, "I think that's it. Okay?"

Farmer agrees: "Yeah. Let's do a take on *Embraceable You*—all right?"

Almost immediately, the tape is rolling, and they are into the piece. It is beautiful from the first notes: Farmer's opening solo, the sensitive interplay with which Hall supports him, Hall's own passages, Swallow as sensitive as Hall under Farmer's later chorus.

At the end there is a moment of silence as the studio reverberations whisp off. Someone says quietly, "Wow!" Then Ertegun's voice comes simply over the speaker, "A masterpiece."

The rest is anticlimax: Hall is unplugging his amplifier, Perkins is packing his cymbals, and Swallow is slipping the cover over his bass.

As all this is being done, Ertegun says, "Art, listen to the last take of *Loads of Love* again."

But Dowd, before getting to a playback of *Loads of Love,* reruns the tape of *Embraceable You*. Nobody minds. And one has the feeling that the reputation of these players and their group might stand by that performance alone. (*1963*)

Blues Night

By 8:30 p.m. the vibrant sound of live music can be heard at the top of a wide staircase on 125th Street near Fifth Avenue that leads down to the Celebrity Club. The piece is a medium blues with the strong flavor of the Southwest and Kansas City, circa 1938.

At the bottom of the stairway, to the right, at the entrance to the club, there is a table that holds a couple of stacks of tickets and a change box. It is recognizably presided over by Victoria Spivey. She has on a tarnished-gold semiformal dress, a matching stole, a pair of high-heel boots, also gold; and she is sporting a new, short-clipped hairdo. The printed tickets on the table in front of her read TRIBUTE TO THE GREAT PIONEER OF THE BLUES, MAMIE SMITH, TO HELP RAISE FUNDS FOR A MONUMENT TO HER ILLUSTRIOUS MEMORY.

Earlier handbills and announcements for the event had said that the evening would feature, among others, Jimmy Rushing, Lucille Hegamin, Hannah Sylvester, Blue Lu Barker, Lillyn Brown, Rosa Henderson, Sam Theard—a heady history of early vocal blues recording and Negro-American cabaret is implied with those names.

Victoria Spivey had a successful blues record in 1929, *Black Snake Blues.* She made it when she was sixteen. She had gone to St. Louis from her home in Dallas, her head full of determination to make a record and also full of the conventional warnings about what can happen to unwary young girls at the hands of big-city slickers. When she got there, she walked boldly into the Okeh recording studios, demonstrated her singing and her piano, and soon had a record date for herself.

Miss Spivey also was a leading actress in an early and still celebrated sound film, *Hallelujah,* directed by King Vidor. In good Hollywood fashion, she did not sing in the movie, although almost everyone else in the cast did. She continued recording and singing into the '30s and '40s. She and trumpeter-singer Red Allen, then she and guitarist-singer Lonnie Johnson, were successful recording teams for a while. Recently, she has become more active again and made new recordings for Bluesville and for her own Spivey label.

And Mamie Smith. Mamie Smith was the first woman to

record a vocal blues. She did so in 1920—*Crazy Blues*. It was an instant success, reportedly selling 75,000 copies in its first month, a phenomenal sale in those days, and it established the recording of blues song for once and all, the line of descent unbroken to this day.

Not that *Crazy Blues* was actually a regular 12-bar blues, and not that it "made" Mamie Smith; she was a highly successful and well-paid performer during the teens of this century. She carried with her a group, Mamie Smith's Jazz Hounds, which, at various times, included such jazzmen as pianist Willie (The Lion) Smith, trumpeters Bubber Miley and Johnny Dunn, tenor saxophonist Coleman Hawkins, and clarinetists Garvin Bushell and Buster Bailey. But when Mamie Smith died in 1946, she was penniless, reportedly deprived of her money by managers and hangers-on. She was buried in an unmarked, triple grave in a Staten Island, New York, cemetery.

The female singers on tonight's guest list—Lucille Hegamin, Hannah Sylvester, Lillyn Brown, and the others—had been able to record their blues, too, once Mamie Smith had made the way.

The Celebrity Club is a large, basement room, with a bar the length of its back wall. The main area of the club is taken up by tables and a good-sized dance floor. Against the wall opposite the bar, there is a bandstand that also is good-sized. The side walls are painted with woodland scenes dominated by birch trees, and the various pillars that reach from ceiling to floor around the room are covered in a cloth that looks like birch bark. The lighting in the club is a fairly dim amber or red. The public-address system seems to have its somewhat harsh speakers planted everywhere.

The music one is hearing is indeed Southwestern, circa 1938, for it is provided by Buddy Tate's band, a small but often highly spirited ensemble employing two reeds, trumpet, and trombone, plus guitar, electric bass, and drums.

Tate is a fixture at the club. When he isn't playing conventional ballads from conventional stock arrangements for conventional dancing, he may take out his own book of originals and play a blues or a medium- or up-tempo jump tune. Most of these are in good Southwestern style, an honorable and, with Tate's group, still robust tradition. Some of these arrangements also make effective use of the band, getting a full sound from a spare instrumentation. This evening, however, Tate is in Toronto with his fellow ex-Basieite, trumpeter Buck Clayton. Reed man Rudy Rutherford stands in his place on the Celebrity Club bandstand.

By this early hour, there are about fifty people scattered around the room at tables. Most of them are the middle-aged jazz fans of New York. And several are long-standing record collectors. Some are thereby staggeringly erudite in discography, and they do invaluable work in the field. Some are deeply responsive to music. But others are jazz antiquarians—or they are simply record antiquarians, perhaps as interested in master and take numbers as they are in vocal or instrumental numbers.

At the door, Mr. and Mrs. Zutty Singleton arrive, greet Miss Spivey, and take their places with a group of friends at one of the larger tables. Trombonist Dickie Wells enters, looking quite young this evening. At almost every entrance there are shouted greetings that burst across the room, accompanied by robust waves of arms and underlined by warm chuckles.

As *Star Dust* finishes, Wells takes his place on the stand beside the band's trombonist, Eli Robinson. Ah, he is here to play! Rutherford signals *"A" Train,* and Wells begins in his really beautiful but firm lyricism.

No one seems to be paying much attention to the music

now. The audience is waiting for the blues queens and kings, who are seated to the left of the bandstand at a long table, a row of gray heads and interesting, lively faces, nodding now and then to the music and chatting quietly. Nearly everyone in the room glances at the long table from time to time, in curiosity and expectation.

Directly across from the guests of honor, on the opposite side of the bandstand, there is a young man fascinated not only by them but by the fact that he is in New York listening to live jazz played by American musicians. His name is Karlheinz Kesten, a pianist and the secretary of the Hot Club of Iserlohn, Germany. His presence has a lot to do with the reason for this unusual benefit.

Last fall a group of American blues singers visited Germany on a tour arranged by producer and critic Horst Lippmann. They included Miss Spivey, Rushing, Big Joe Williams, Muddy Waters, and Lonnie Johnson, among others. Miss Spivey met Gunter Boas and his wife, Lore, a German couple whose interest in the blues goes back many years and who are members of the Hot Club of Iserlohn.

Inevitably, the three spoke of Mamie Smith. The Boases were shocked to discover she was buried in an unmarked grave. They wanted to do something about it. In late November they held a benefit concert in Germany with six groups, raised some money, and obtained a tombstone for Miss Smith. The Hamburg-American Steamship Line agreed to transport the stone to the States free of charge, and appropriately they used the *SS Iserlohn*. There was a ceremony when the ship departed, including music by musician members of the Iserlohn club. Kesten made the trip with the headstone.

The Iserlohn docked at New Orleans, and although customs allowed the stone to pass through free, there was a charge for duty, a charge for express to get it to New York City, and a charge, the highest of all, to have Mamie Smith's remains moved from the three-level grave to a plot of her own. Miss Spivey and Lennie Kunstadt decided to hold a benefit on this side of the water to try to raise the additional money, and the Celebrity Club management donated a Monday evening.

There are now about one hundred people at the tables and bar, and there are new arrivals on the bandstand too. Trumpeter Pat Jenkins takes his place as clarinetist Tony Parenti arrives, shouts a greeting toward the bandstand, and takes his seat with a couple of friends at a nearby table. The rumor that Rushing is here hurries across the room.

The band is into a medium blues again. Jenkins finishes his solo, and then it is Rutherford. He stands there, knees bent, in a neat, conservative blue suit, looking as polished and as prosperous and almost as complacent as a stockbroker who lives quietly in the suburbs and seldom goes out in the evening. Yet he is saying things with his clarinet that are full of terror and love and joy and beauty. That is the way of the blues if you can play them, and Rutherford can play them. Wells musically instructs the band, setting a riff figure behind Rutherford, and the rest of the players fall in immediately.

It is 10:20 now, and suddenly a familiar voice fills the room, ringing through the loud-speakers: "I want a little girl. . . ." The crowd looks up, and there is Rushing, standing in front of the bandstand, holding a hand microphone, a slight, sly smile on his face. He is a big presence of warm and easy charm. Wells starts to improvise behind him now, and suddenly it is 1939, not in a nostalgic and half-realized echo but almost in reality—one half expects to hear a Lester Young solo at the end of Rushing's chorus. As he finishes, there are shouts and loud applause. And then, of course, he goes into a blues: "She's little and low and built up from the ground. . . ."

Seven choruses later the applause is louder still, and then the emcee of the evening, Boots Marshall, is at the microphone proclaiming, "We want you to enjoy yourselves and make yourselves at home; we've got a lot of stars to come. . . ."

At the back of the room Horst Lippman has entered with his friend, guitarist Attila Zoller. They take their places at the bar, where they are greeted by Kunstadt and introduced to several patrons.

The band begins another instrumental blues on a heavy boogie-woogie bass figure. Rutherford is playing clarinet now; he has a big tone. There are three couples dancing, executing a kind of becalmed, businessman's Lindy.

"I came to sign up some people for another blues festival for next year in Germany," Lippmann is explaining to a new acquaintance at the bar, as a sudden burst of laughter rises from a nearby back table.

Danny Barker is at the entrance now, his guitar in one hand and his wife, Blue Lu Barker, by his side.

"Danny! Man! What on earth. . . . Where did you get that hair? When did you start wearing that thing?" There is broad laughter from Barker and a friend at the bar over the toupee he is sporting, and Barker protests innocently about "my rug."

It is obviously going to be a long evening, and a couple of unhardy souls give up and head for the front door.

At the bar, several heads turn to note the somewhat unexpected presence of composer-pianist Tadd Dameron.

The band has reassembled now, and, with all the sitting-in, it is almost a different group—it is certainly a larger one. The players are laying down some Kansas City-style riffs that probably remind Dameron of his youth and his days with the Harlan Leonard Band.

As the piece ends, emcee Marshall starts acclaiming "that wonderful lady who is responsible for what goes on here tonight, Miss Victoria Spivey. . . ." She marches forward to acknowledge the applause. At least half the room expects her to sing, but suddenly she is gone, and Marshall is doing a fast and loud *Just One of Those Things.* ("A trip to the moon on gossamer wings" on *blues* night?)

Dameron has slipped in at the piano now, and Marshall begins to introduce the guests of the evening at the table of honor to his right. "A young lady that can still make high C, Miss Rosa Henderson." She bows from her seat at the table. So does Hannah Sylvester, another who first recorded her blues in the '20s. The crowd is surprised at this, for most of the audience had expected her to sing. Miss Sylvester had recorded again recently, after all, for Bluesville and Spivey records.

Then Lillyn Brown, a woman of nearly-white-haired dignity is at the microphone.

"You won't believe this," remarks the venerable actor Leigh Whipper to a young acquaintance standing in the rear of the hall, "but she is seventy-nine years old."

Miss Brown is speaking with clear and fluent energy about how she came to write one of her numbers "a few years ago," and then she is into it, *I'm Blue and Rockin'*. Her voice is as big and ringingly precise as that of a woman of thirty, with no aging elderly vibrato or cloudiness. And her musical drive matches its clarity.

As the audience shows its delight after her last verse, Miss Brown seems equally delighted. "I have a little short one now," she says. She was one of the first to follow Mamie Smith on records, but she is not trading on her past tonight—she is singing here and now:

If you want me to love you,
Please don't make me cry.
If you want me to love you,
Please don't make me cry.
'Cause if you make me cry, baby,
My love just seems to die.

As she leaves to return to her table, the audience is again smiling broadly over its applause. And the musicians are smiling perhaps broadest of all.

"Look," says a middle-aged fan at a side table, "that's Louis Metcalf on trumpet sitting in now."

The continuity of guest performers continues. Blue Lu Barker is standing front and center of the bandstand in a blue dress. Danny Barker, looking rather mild in contrast, is on her left, his right foot propped on a chair and his guitar resting across his knee. Suddenly neither of the Barkers seem mild, for they have gone into *Hot Dog! That Made Him Mad!*

During the applause at the end, half the crowd seems humorously bracing itself for their best-known number—and they anticipate correctly. It is, according to Danny's announcement "by very special request," *Don't You Feel My Leg*. Broad laughter shatters Lu's opening verses.

Then comes Sam Theard ("I'm a sick comic, you know—sick of being out of work"), who has been doing comedy and songs at least since the '20s, when he started in his native New Orleans, but who is probably best known for having written *I'll Be Glad When You're Dead, You Rascal You* and *Let the Good Times Roll.*

"Gimme that E-flat arpeggio," he is tossing over his shoulder in Dameron's general direction. "Ah, that's nice. Do it again."

"I'll be glad when you're dead, you rascal you . . ." he

sings, and then goes into an energetic, acrobatic dance across the floor in movements that seem unlikely in the neat, tan, tweed suit he is wearing.

Then clarinetist Parenti. By invitation he borrows Rutherford's instrument to play a slow clarinet blues for the crowd.

"How does he sound to you, compared to me?" Rutherford asks a fan standing to the left of the dance floor.

"Your sound is fuller, I think. But, you know, all of those New Orleans clarinetists have a certain lyric thing. I don't know how to describe it, exactly, but they all have it."

"Well, you see I asked you because he is using my instrument. But I learned from him too, so. . . ."

Parenti's sound is still billowing across the room as a somewhat disappointed patron consults Miss Spivey near the door. "Victoria, Queen, I didn't come here just to watch you take up money for Mamie Smith or take bows," he says with a sort of half-smile. "I came here to hear you sing."

"Well." She looks up, pausing. "My guests come first." Then with a slight laugh: "And we're so far behind now, I may not get to sing at that."

It was nearly 2:00 a.m. when Miss Spivey finally did sing. By then she had passed out about 200 admission tickets at the door.

John Bubbles of the old Buck and Bubbles vaudeville team had come by, spoken a song in his recitative style, and danced charmingly.

Lucille Hegamin, the first to record a blues after Mamie Smith, had sung *He May Be Your Man* (*But He Comes to See Me Sometimes*).

Maxine Sullivan had arrived and had sung her extended version of *St. Louis Blues* with a voice still sounding like 1940 and with several encouragements to the band members to solo between her verses.

Rushing had come back.

And others had sung and reminisced.

When Victoria Spivey stood up at the mike, she did two numbers, and one of them was *Black Snake Blues*. As she started it, about half the room probably felt that, for the time being at least, all was right with the world. (*1964*)

Bash It

The Impact Studios are on West Sixty-fifth Street, and they are, therefore, in what is now called the Lincoln Center area of New York City. But Impact's narrow building was there long before Lincoln Center was a gleam in a real-estate speculator's eye or a twitch in a culture-monger's pocketbook.

Impact is a second-floor walk-up, and en route one might assume that its building is otherwise deserted. The studio is a favorite of Don Schlitten, jazz producer for Prestige records, and Schlitten had booked it for 1:00 p.m. on a day this spring to record pianist Jaki Byard "with strings." The "strings" were to be Ray Nance's violin, Ron Carter's cello, Richard Davis's bass, and George Benson's guitar. Also set was Prestige's favorite drummer, Alan Dawson, doubling on vibraharp. There were to be some formal arrangements by Byard and some head arrangements to be worked up on the date.

By 12:55, Carter, Davis, and Benson, with the help of Impact's Eddie Heath, were set up inside the studio. The small room now seemed a mass of waist-high baffles and sounding boards and an intricate tangle of stereo microphones, booms, and wires. These somehow managed to leave enough space for a piano, drums, vibes and strings, and perhaps even the musicians to go with them.

Benson entered the control room to the rear of the studio and, spotting a particular mixer against the left wall, remarked that this was the kind he intended to add to his own equipment. Within a few minutes, Nance had arrived and greeted the other players. He quickly had his violin out, took a look at his music, and asked Heath for a music stand.

Impact's engineer, Richard Alderson, was checking dials and tapes when Byard and Dawson entered at 1:10. Byard announced himself to one and all with a broad, mock-serious, "Well!"

He was dressed in a tweed jacket and flannel trousers, and he wore a tie.

He entered the booth to chat with Schlitten and then soon got down to business. His voice conveyed a combination of energy and eagerness, with a bit of nervousness.

"I'm going to use *Exactly Like You* to jam a little," he said.

"Why don't you play *Take the 'A' Train* at the same time?" Schlitten remarked, proposing a counterpoint of two familiar tunes built on closely allied chord changes.

"Oh, I've got something better than that," Byard answered cryptically.

Like most jazz musicians, Jaki Byard is a worshipful admirer of the late Art Tatum. Furthermore, Byard has the facility and dexterity as a pianist to reproduce the master's style, but with a strength and touch very much his own. Such an accomplishment might be enough for most men, but Byard is also a thoroughly professional alto and tenor saxophonist, and a capable trombonist, trumpeter, guitarist, vibist and drummer. He is, further, his own composer and arranger.

As if it were not enough, Byard is virtually a functional history of jazz piano: without patronizing his sources or compromising his own identity, he can perform convincingly in almost any idiom from the Harlem stride of the 1920s through the innovative, "free form" improvisations of the 1960s.

Informing all of this is a quick, flexible sense of comedy which enables Byard to spoof without ridicule—enables him, for example, to juxtapose a Fats Waller-style bass line with a Bud Powell-inspired treble melody, with respectful, enlightening, and hilarious results. Indicating that he does not consider his musical humor merely easy joking, Byard once told Dan Morgenstern, "I don't play tongue-in-cheek and I hate to hear people say this about me. I think hardly any of us can be completely ourselves, pianistically, what with all the people that have been before us, so I try to go into each phase of the piano with respect. If you're going to do it, do it all the way."

Byard's humor is frequently unexpected. I have seen him, under the pressures of a time-limited recording session seem strictly (and nervously) business, but get off hilarious side-remarks along the way, sometimes verbally, and sometimes from his keyboard during a run-through.

Most of the foregoing has been known about Byard to many musicians and some followers of jazz, particularly it was known to musicians and followers from the Boston area, for over fifteen years. It is only in the past four years or so that the rest of the country, and the world, has caught up. But now the once underground reputation of Jaki Byard has come overground.

John A. "Jaki" Byard, Jr., was born in Worcester, Massachusetts, in 1922. His father played baritone horn, his mother played piano, and he himself was sent for keyboard lessons at eight. His lady teacher taught him to reproduce such popular classics as *The Scarf Dance* and *Humoresque* almost by rote, and when the Depression hit the Byard household and the lessons ended, young Jaki was left with little or no theoretical background to fall back on.

From the ages of ten to sixteen, Byard spent less time playing than listening. He listened to any band that appeared locally, and he heard some of the best: Fats Waller, Fletcher Henderson (Coleman Hawkins was in the ensemble), Count Basie, Joe Venuti, Isham Jones, Chick Webb with Ella Fitzgerald, and the most intriguing of all the bands to young Byard, that of Earl Hines. Further, "my old man hipped me to Teddy Wilson," he explains. "He said that that style was going to be the big thing."

His father also had a trumpet which he passed on to his son. The trumpeters to emulate then were Roy Eldridge and Hines's Walter Fuller, and Byard emulated. But it was on piano that he began to gig locally at sixteen with the bands of Doc Kentross and Freddy Bates's Nighthawks. His very first job, with Kentross, was for a high school dance, and was so important to the youngster that he played it with a badly hurt left hand in which six stitches had been taken that afternoon.

A friend from this period, novelist Don Asher, remembers that Byard practiced late at night when he could borrow the use of the piano in the basement of the local Jewish Temple. And Byard, in turn, remembers giving fifty-cent piano lessons to Asher, who still, by the way, plays professionally in San Francisco.

Byard soon began an apprenticeship in harmony with local guitarist Lenny Waterman, and did his first arranging by scoring Coleman Hawkins's celebrated *Body and Soul* solo for Freddy Bates's band. He returned seriously to piano study at this point, with Lennie Sachs in Boston, when he was drafted in 1941. An army career that started badly for

Byard took a decided turn for the better when he was put in a barracks with drummer Kenny Clarke and "a fine pianist," Ernie Washington. On duty, Byard took up trombone and Washington struck glockenspiel in the marching band; in private they had piano sessions and Washington became a major influence.

Back in Boston after Army discharge, Byard became entirely serious about music study. Unable to attend school, he discovered the schools' teaching methods, then attended the local libraries, and taught himself according to the same plan. "But don't say I'm self-taught," he insists. "I studied with other people, and we'd get together and discuss things . . . Schoenberg also studied in the library, but would you call him self-taught?"

Meanwhile, Byard was earning his way, paying his dues, with the local Boston bands like those of Dean Earle, Hilary Rose, Danny Porter, and Phil Scott. It was during a two-year stay with the group of violinist Ray Perry that he was encouraged to double on tenor saxophone. ("Play as many as you can," Perry told him. "It's good for you. Look at Ray Nance and Benny Carter.")

In 1947, Byard went on the road as pianist with the very popular group of alto saxophonist Earl Bostic, and at this period he also discovered others of his major influences: Erroll Garner, Charlie Parker, and Bud Powell, a fact which did not always please his leader ("my guys on piano played behind the beat, while Bostic liked to go forward").

Back in Boston a year or so later, Byard formed a rehearsal band which included trumpeter Joe Gordon and saxophonist Sam Rivers. "We were attempting to be the hippest bebop band in town." There followed a year on the road with a "pretty bad" touring stage show, but a band that included Thad Jones, Gigi Gryce, and Jimmy Crawford.

It was on his next return to the Boston area that Byard's reputation really began to build. He held forth for three years at the Melody Lounge in Lynn with Charlie Mariano, an alto saxophonist, and Dick Wetmore on trumpet and violin. Byard himself came to terms with Art Tatum. He had heard him before when he was in his own Hines-Waller period and did not then fully discern how much Tatum had gone beyond Waller. This time Byard was floored, as he puts it, by Tatum's work and at first temporarily retreated to his saxes.

On tenor, he joined trumpeter Herb Pomeroy's Jazz Workshop band, sitting in the sax section with baritonist Serge Chaloff, and also adding arrangements to the book. "*That* was the one they should have recorded," Byard says. It was "the most fiery band you'd ever want to hear."

On his own again, and back on piano, Byard found a group of younger musicians gathering around him, including trumpeter Don Ellis, drummer Al Francis, and pianist Dick Twardzik. Without being academic about it, Byard and his associates wanted to do something in jazz beyond what had ben done before. ("I hate to hear the same thing over and over" is a favorite Byard admission.) And in 1960, when Byard and Ellis both joined Maynard Ferguson, something of a schism developed within the band among those members who stood for jazz as it was, and players like Byard and Ellis, who favored experiments with unusual time-signature, unusual harmonic understructures, or even off-the-wall improvisations without regard to pre-set thematic or harmonic outlines. Ferguson recorded Byard's *Lucky Day* and *Extreme,* but Byard did less arranging for the trumpeter than he would like to have because of arranging competition from another band member whom Jaki likes to refer to as "the B-flat king."

With such his commitment to the future of jazz, it was perhaps inevitable that Byard would encounter the very important group that bassist Charlie Mingus led at the Showplace in New York, with the late alto saxophonist Eric Dolphy, who became one of Byard's favorite musicians ("the direction the alto is going in reality . . ."). Before long, Byard was himself working with Mingus both as pianist and arranger, and was associated with Dolphy in person and on recordings.

But Jaki Byard, the underground musician, was ready to emerge fully. He left Mingus in 1965 and has been on his own since, leading his own small ensembles.

Something of the generous, comprehensive respect with which Jaki Byard holds his heritage is indicated in his remarks on others. For example, "one of my favorite sax players was Chu Berry. I often wonder what would be the condition of jazz if cats like Chu, Herschel Evans, Fats, Bix were living today. I used to adore Willie Smith (the alto saxophonist). With Lunceford he played some fantastic things." Or, "James P. (Johnson) was a fine composer, and his longer works deserve more attention than they've received so far." Or, "Billy Strayhorn's sure a beautiful cat, isn't he?" Or, "Ben (Webster), that was my stick on ballads . . . that beautiful sound." Or, "For me the most astounding thing about (Thelonious) Monk as a composer is his lyric sense . . . so strong that you can sing anything he writes; and always, you know it's *him*."

And of himself? His grand stylistic versatility is "not a gimmick thing," not stunt work, nor a series of quick imitations. However, as he once told David Himmelstein, "I can't sit there and just play single lines all night and go away satisfied. I can't play one way all night; I wouldn't want to and I wouldn't want the public to hear me that way. If you stay in one groove, you can't reach the people."

Back in the Impact Studio, Richard Alderson had moved a few mikes around and formed Ron Carter, George Benson, and Richard Davis into a kind of ritual circle of strings. Carter began a riff, and Nance and Davis soon joined him. They were running over the introduction to the first piece.

"Fellows, make those even eighth-notes," said Byard entering the studio. His voice was authoritative and friendly but somewhat edgy. Then he tossed over his shoulder to Don Schlitten, "It'll be good when I get in there and play with them."

He crossed to the piano and said, "Okay, let's try the intro to the first piece."

It turned out to be the *Girl Watchers* theme—from television commercial to pop hit to jazz vehicle—and the most difficult arrangement of the day.

"Try it again," Byard said. And to Dawson, "Bash it, Alan."

At the end, Byard stood, moved toward the other players and said, "Good! How'd it feel?"

"I can't play it that fast," Nance protested.

"Okay, *he's* got the changes," Byard said, pointing to Benson. "Just do something lyric on the changes."

Byard rehearsed the piece section by section. The introduction. The first chorus. The fugal interlude, which featured Carter and Ray Nance. The ending. As usual, the individual musical parts had a few copyists' errors and some occasional dubious notes.

"Four before C," Carter said at one point. "Is that note B flat or A natural?"

After a run-through, Byard had some new ideas on his score and asked, "Want to do me a favor? Repeat those two bars there. I want to vamp that. But make that an E flat."

"Of course, I can read," someone was protesting a moment later. "But not on Tuesdays. This is my day off."

"We will be here," Schlitten emphasized behind the glass, "until we have finished the album."

Perhaps it was time to change the subject. Perhaps the

musicians needed to relax a bit. At least that seemed Byard's idea when he asked the room in general, "What do you all want to jam on?"

There was no response. The musicians seemed to want to get the *Girl Watchers* right after all, and they went back to it. Byard omitted one section that wasn't sounding too good anyway and announced, "We've got five more minutes, and we're going to record it."

"Famous last words," someone muttered.

"Well, don't worry about the fugue—we'll do that and splice it in later," Byard said.

Twice more on a difficult section and Byard jumped up, announcing, "Good!"

"But how much do I play?" Benson wanted to know about his solo.

"For one whole chorus after the *DE doten de da, DE doten de da.*"

After another run-through, Byard changed his mind, saying, "Good! Take another one, George. Make it two for you."

Carter had an idea. "If we could cut out some of those notes in there," he said, pointing to his score, "it would go a little smoother."

Byard's response was aesthetic practicality itself: "Play what you want to."

Alderson entered the studio, moved Benson's amplifier and shifted Nance's chair and mike. He seemed satisfied, and that meant that everyone was now ready to record.

"Stand by," Schlitten announced on the tape. *"Girl Watchers Theme,* Take 1."

Immediately, the players got busy, and their energy filled the studio. Nance bowed at that odd angle he uses. Carter looked studious. Davis looked dead-pan, but as if he might offer a hilarious musical joke at any moment.

About halfway through, Byard stopped the take, announcing that he goofed, and that, anyway, they had better hear a playback in the studio for balance. "In here, I can't hear the cello or guitar."

Ten minutes later, Carter stopped the second take, waving his bow in the air and announcing, "I goofed." But it was a sort of ensemble goof only a discerning ear could have heard.

"Would it help you, Ray, if I give you an amplifier in the studio?" Alderson wanted to know. Nance said it would, and as Alderson installed the amplifier, Carter and Benson strummed a background to their own quiet conversation.

Take 3 was a false start, and to begin the 4th, Byard counted off firmly "1, 2, 3, 4," onto the tape. Benson threw himself into his solo, eyes tightly closed, forehead occasionally in a momentary, intense frown that seemed to spur him on.

"Good—do another one!" Byard encouraged quickly at the end. Nance seemed to sprawl in his chair as he played, but he was obviously in control of what he was doing. Automatically, he dried his left hand along the length of his trousers at every rest.

One final version of the introduction to splice in the beginning, and Byard was satisfied.

"Good, thank you, fellows," he said, and because he knew it was time for some solos all around, he announced, "Now. Everybody know *Exactly Like You?*" There was general nodding or silent assent around the studio. But Byard had more up his sleeve. "Ray, you play *Jersey Bounce.* You know that? George, you do *Darktown Strutter's Ball.* I know you're not that old, but you might know it. Richard, you know—what's that thing (he hummed a few notes)—Stan Kenton isn't it?"

"Intermission Riff," somebody called from across the room.

"Alan, you play *Ring Dem Bells.* Ron, you play that bossa nova thing (again he hummed a couple of bars)."

"Desafinado," someone offered.

"And I'm going to play *'A' Train,"* Byard said. "I'll play the bridge." And then, "These things all have the same changes," he added as an afterthought. "Ray, take the first chorus, Ron, take the second, George. . . ." Byard continued to assign the order of improvising.

In a few minutes, he had passed out a written introduction, and Schlitten was announcing onto the tape, *"Exactly Like What,* Take 1."

A couple of run-throughs and false starts later, they were really into it. The introduction, intriguingly oblique, skittered by. The opening ensemble, which promised melodic chaos, proved to be complex but lucid. Nance set a groove in his opening solo, and Davis, as his sole accompanist, responded. When Benson came surging in for his solo, Davis was joined by first Carter and then Dawson. Each man took two choruses. Byard's energy was climactic. It was going to work.

"What is Jaki Byard going to do next?" Alderson wondered admiringly in the booth. "You never know."

Byard was not quite through with this arrangement. "For the last chorus," he announced, moving from his keyboard to the center of the strings, "everybody play harmonics in C. Al, play up high. And everybody go for himself on the bridge. Try it."

They did, and the effect was stunning—the only way to end a performance that had begun like this one.

"How would it be," Nance asked before a take, "if I bow the opening? Because I'm having trouble playing pizzicato, and I'm not with them anyway."

"Try it," Byard offered and, at the end said, "Thanks a lot, Ray. That was a good suggestion."

On the final take, Nance again took the opening solo, and again his superb swing set the groove. The groove set, he began to soar, and the others soared with him. The ending still worked.

Just before the playback Byard announced, "Take a break, gentlemen. But be taking a look at this ballad."

He entered the engineer's booth to hear the playback, and after he had confirmed his satisfaction with *Exactly Like What,* Byard re-entered the studio with a broad smile and said, "Well, we'll continue making history."

The next document for history was the date's ballad, Byard's *The Falling Rains of Life.* By 4:50, he was announcing, "Okay, gentlemen, here we go for a run-through." The leader had the gentle theme on piano, Nance an obbligato, Davis a mysterious, double-time walk.

"Improvise," Ron!" Byard shouted at one point.

At the end, Nance joked, "I ran out of music before the rest of you did."

Byard again left his keyboard to speak to the strings: "It's perfect the way you all did it. After this, don't worry. We'll just play for Alan's vibes solo. But it needs *passion!*" He gave some mock-serious gestures with his arms and then squeezed his fists under his chin and added, *"Romanticism!"*

After another run-through, Schlitten entered the studio. "What's wrong?" Byard asked.

"Well, it's fine but it's getting a little long," the producer said.

By the time the second take was going onto the rolling tape, Dawson had a half-chorus, and Byard's sustained ending was enhanced by Nance, who knew just when to enter on top of it and just how much to play when he got there.

By 5:20, Byard had distributed a new piece among the

players. By 5:35, he had expressed dissatisfaction with the way it was going, and he took up the music parts. He turned to Schlitten behind the glass panel and invited suggestions.

Schlitten entered the studio and said, "Let's jam *How High the Moon* and begin it as a ballad for Ray. Later, how about some fours between Ron and Richard?"

Byard and the other players seemed to agree.

"That's what jazz is all about anyhow," said Schlitten, re-entering the booth. (*1968*)

IV

ANNOTATIONS

*Count Basie in Kansas City: The Benny Moten Orchestra, 1930-32**

In mid-December of 1932, in the depths of the Depression, a group of musicians entered the Victor recording studios in Camden, New Jersey, for a marathon recording session. They were the members of the Bennie Moten orchestra, one of the leading bands of the Midwest—of the country, for that matter—an organization which had been recording for Victor since 1926 and for another company three years before that. The men were demoralized, literally hungry, and the long session was almost the last act of this particular manifestation of the Moten band. As clarinet and alto soloist, Eddie Barefield has put it, "We didn't have any money . . . we had to get to Camden to record, and along comes this little guy Archie with a raggedy old bus, and he took us there. He got us a rabbit and four loaves of bread, and we cooked rabbit stew right on a pool table. That kept us from starving, and then we went on to make the records. Eddie Durham was doing most of Bennie's writing then; I made *Toby* that time. We just turned around and made it back to Kansas City. We hung around there for a while, not doing much of anything . . ."

Certainly there is nothing either in the music or in the frequently joyous way it's played on this LP to indicate that the band was in such straits at the time. Or glance at the

* *"Count Basie in Kansas City: The Benny Moten Orchestra, 1930–32,"* RCA Vintage LPV-514, reissued as RCA AFMI-5180. Reprinted with permission.

soon-to-be-illustrious personnel. There is "Hot Lips" Page as the group's trumpet soloist; there are Dan Minor and Eddie Durham as trombone soloists, with Durham also contributing guitar solos and arrangements; there is Barefield; there is Ben Webster on tenor saxophone; there is William, later "Count," Basie on piano; and there is Walter Page, whose firm, steady but exhilarating four-beats on bass had as much to do with the character of this music, probably, as anything else. And there is a style of music which is fresh and rather unlike Moten's previous music and which, in just a few years, would dominate big band jazz.

But in late 1932, apparently few people wanted to hear it. They wanted to hear earlier Moten instrumentals like *Moten Stomp* (which sounds quite dated now), and they wanted to hear *South* (which Moten had first recorded in 1924, and which in its 1928 version could be heard in urban jukeboxes well into the early forties). But they didn't want to hear this music which sounds so frequently vital and undated.

We have kept that 1932 session intact and in the order of the original recording, omitting only two selections outright—one almost entirely taken over by the Sterling Russell vocal trio, the other to a vocal by Josephine Garrison. The first piece, Barefield's *Toby,* introduces the major soloists, Durham, Page, Webster (notice how the riff that accompanies him becomes the next chorus, transferred to a brass lead), Basie, Barefield, and (in my opinion) Minor on the brief open trombone solo and Durham on the plunger solo. These surging, inventive, shouting brass and reed figures here may remind one more of the later Lunceford style than the later Basie style. The performance probably holds a special meaning for those who were lucky enough to hear the band when, as Barefield puts it, "Lips Page would play maybe fifty choruses and we would make up a different riff behind each chorus."

Moten's Swing succinctly reveals what this band achieved. Here, by late 1932, was a large jazz orchestra which could *swing* cleanly and precisely according to the manner of Louis Armstrong—a group which had grasped his innovative ideas of jazz rhythm and had realized and developed them in an ensemble style. Further, the piece features original melodies on a rather sophisticated chord structure borrowed from a standard popular song. Notice also how much Fats Waller there is in Basie's solo—the stride masters James P. Johnson and Luckey Roberts were Basie influences in his early days, and Waller gave Basie instruction. Notice also how personally melodious "Lips" Page is. And notice that the figures usually played today as *Moten's Swing* are only the final riffs in this performance.

Blue Room opens in a somewhat older style, but there are the chimes Basie provides behind "Lips." Then, hear the way the spurting riff figures begin to appear under Barefield's wry solo. Finally comes the thrilling brass and reed riffing, which not only takes us into the later world of Count Basie, but which in itself is one of the most beautifully played passages in all recorded jazz.

New Orleans was Jimmy Rushing's vocal for this session, but it also has a wonderful, eruptive moment on the verse of the piece by Ben Webster who sounds a lot less like Coleman Hawkins here than he is supposed to have sounded at this early point in his career.

The Only Girl I Ever Loved contains a rather square style vocal by the Sterling Russell Trio, but the side is notable as a sample of the group performing at slower tempo. There are also brief, effective flashes of Durham's guitar; there is Ben Webster; and there is an ending which, in effect, reinterprets the piece as if it were *Moten's Swing*.

Milenberg Joys, like *Prince of Wales,* reinterprets a piece from the earlier jazz repertoire in the new style. Basie is lightly humorous in his deliberately old-timey introduction, and do not miss the saxophone figures, nor the way the rhythm walks behind them, nor the chase between Barefield and Webster toward the end.

Lafayette is one of Durham's contributions to the new repertoire, not quite so lean and spare in the writing as his later arrangements for Basie, however.

I am sure that *Prince of Wales* would be an exceptional performance if only for the forceful and handsome, not to say joyous, striding of Basie's piano—we tend to forget how beautifully he could play this style and what glee there was in it. But also do not miss the way Walter Page comes in behind Webster, the figures behind "Lips," and the way that the ending in effect rewrites *Prince of Wales* as if it were *Toby*.

Walter Page was the solid foundation on which this band was built rhythmically. And Walter Page's Blue Devils, his Midwest group of a few years earlier, was the source of its style and of many of its best sidemen, beginning with Basie and Jimmy Rushing, then "Lips" Page, and ending with Walter Page himself. It is fascinating and instructive to see how the Moten band evolved as these men gradually joined it. Indeed, the earlier Moten records which sound best today are apt to be those on which the ex-Page men appear and contribute.

The Jones Law Blues comes from Basie's first recording session with the Moten band, and it features a brief Basie solo. But the style is older and shows one of the Moten band's earliest debts, for it might almost be one of Jelly Roll Morton's late RCA Victor records. *Small Black* is more typical of the late-twenties Moten—"peppy" rather than really swinging, with solo moments from Ed Lewis, Bus Moten, Durham, Woody Walder, and (I think) Harlan Leonard on clarinet. For this version, we have expanded Basie's interesting solo space by first using the chorus from an earlier take and then including the solo originally heard in this take. The first shows Basie's other major influence after Waller, Earl Hines. There is more Waller in the second solo, but there is Hines, too, and the remarkable thing is how well Basie had brought the two styles together as one so that they don't sound like a patchwork.

The *New Vine Street Blues* is a remake of a piece first recorded in 1924—an unusual 24-bar, long-meter structure. It is an altogether remarkable, sustained performance, much less over-arranged and much more emotionally cohesive than most of the Moten records from this period. This, by the way, is a previously unreleased "take," chosen because the solos seem a bit stronger. I assume that the effectively Bechet-like clarinet solo, which has so much to do with developing the mood of this performance, is Woody Walder, who had been with the Moten band on its earliest records, and that the baritone solo, which manages some slap-tonguing without raucousness and without breaking the mood, is Jack Washington. I also take it that the plunger-muted trumpet introduction is Ed Lewis, although this playing is rather different from the Nichols–Beiderbecke style he uses on his other solos here.

This is also a new "take" of *Won't You Be My Baby?*, included for Rushing's marvelous singing—he comes on as if he wrote the piece himself, which, as a matter of fact, he did. Rushing was largely a ballad singer in those days, but here he has the plaintive wail of a blues celebrity of the twenties—that wail plus his own ubiquitous good humor. The performance also shows an effort at a steady four-beat rhythm.

Oh! Eddie is arranged, or over-arranged, in the old style, but it has Basie (it also has Bus Moten's accordion jiving around with the brass), and the last chorus teasingly sketches out ideas that later turned up in *Moten's Swing*.

That Too, Do is a blues, but a blues with a bridge in the

instrumental choruses(!), and Rushing's only recorded blues vocal with Moten. The title is rather like "that fuss" or "that mess," a kind of mock disparagement of the piece (and I'm convinced the title should have been *That To-do*). Notice that each of Rushing's choruses had a life of its own on a later blues with Basie. Also notice that provocative call-and-response between "preacher" Ed Lewis's trumpet and the band's "congregation." (Incidentally, if the bass figure that introduces *That Too, Do* gets under your skin, listen to King Oliver's *Snag It* or to the Modern Jazz Quartet's first recording of *Django*.)

When I'm Alone is more typical of the balladeer Rushing of this period, and the piece shows the tempo at which the four-beat swing of the later band was worked out. Also, watch the hints of things to come in those crisp brass figures at the end.

Somebody Stole My Gal almost had to be included if only for Basie's scat singing. The "jungle" jive that introduces it is perhaps all-of-a-piece with the *vo-de-oh-do* baritone, trumpet, and tenor solos, but Basie's piano contribution to the introduction is a moment of light-hearted joy, and in those brass figures at the end we are, once more, warming up for *Moten's Swing* and jazz to come. (*1965*)

A Celebration of Trumpeters

New Orleans Horns: Freddy Keppard and Tommy Ladnier*

The tradition of New Orleans horns—the singular line of Crescent City cornetists-trumpeters—begins with Charles "Buddy" Bolden and ends with Henry "Red" Allen, Jr.** We have the course of its remarkable development well documented on records, virtually step by step and contribution by contribution.

No, we don't have Bolden's records (or, better to say, we haven't found them yet). And we don't know what Manuel Perez sounded like. But such things being granted, we know a great deal. And from both matters of fact and matters of valid conjecture, we know that in the few records Freddie Keppard made we have moving evidence of what the earliest lead horn sounded like. And in those of Tommy Ladnier, we have a style virtually on the brink of Louis Armstrong's great discoveries about jazz rhythm and jazz horn.

Keppard was the first "King" after Bolden, and he was so crowned by the populace, by the fans. If valid conjecture did not tell us so, then we would still have the testimony of several musicians as to Bolden's influence on Keppard, and to the fact that Keppard's was a more musical, knowledgeable and sophisticated version of the earlier master's style. It was Keppard's music, too, Jelly Roll Morton tells us, that

* *"New Orleans Horns: Freddy Keppard and Tommy Ladnier,"* Milestone MLP 2014. Courtesy of Milestone and Prestige Records.

** Or it did until Wynton Marsalis and Terrence Blanchard brought it back home in the 1980s.

was the basis of the style of the Original Dixieland Jazz Band. It was Keppard who, with bassist Bill Johnson, first took New Orleans jazz out of its home city in 1913 for a tour of the Orpheum Theater circuit.

Tradition has it that Keppard didn't want to record that music with the Creole Jazz Band when he was asked to by Victor Records. That he felt that records would make his tunes and his style too easy for others to steal. Sidney Bechet has put it differently: Keppard felt that records would commercialize a music that better belonged as a mutual communal communication between the players and their audience. (Yet it turns out that he may have recorded in December of 1918 after all; there is a listing in the Victor files for that date of a test record of a piece called *Tack 'em Down* by a "Creole Jass Band.")

Happily for us, in any case, Keppard did record, beginning in 1923 as a sideman with Doc Cook, and subsequently with the pick-up groups represented here.

In Keppard's buoyant, staccato horn, we hear jazz finding its way out of the clipped and limited accents and rhythms of ragtime, as well as its determinedly optimistic emotional outlook. Keppard's great power and technique were reputedly diminished by the time he entered the recording studios (and indeed there admittedly is some fluffing and faltering on his recording of *Adam's Apple* with the group called Jimmy Blythe's Ragamuffins). But the strength and drive of *Stockyard Strut* and the supple ironies of *Salty Dog,* with his own Jazz Cardinals, clearly are the work of a major New Orleans musician. There are two surviving takes of *Salty Dog,* and both are included here, giving us a rare opportunity to hear just how much improvisation, and what kind, Keppard might use. My own preference is for the less familiar first take: it is stronger, meaner, more soulful, more varied—all those things.

The two items here recorded under drummer Jasper Taylor's name were long listed in discographies as featuring Natty Dominique. Clearly it is not he, although I confess that personally I am not fully convinced it is Keppard. There is, in any case, a bit of conventional *vo-de-o-do* in the phrasing on *Stomp Time Blues.* However, that earthy solo on the slower *It Must Be the Blues,* played over a tango beat, is admirable blues horn.

Tommy Ladnier was a limited hornman, limited in range particularly. Yet he was a player of power and drive, and he achieved a striking variety of sonorities, a resource greatly augmented by his use of *wa-wa* muted effects. And he played with deep and moving passion.

Travelin' Blues and *Steppin' on the Blues,* are, for me, two of the most delightful jazz records to come out of the 1920s. Major credit should surely go to Lovie Austin for having produced so much music from a trio of musicians. She employed the particular talents of her hornmen with great perception. (Yes, I do wish this had been the *real* Johnny Dodds, but this is surely the best recording by his reputed rival, clarinetist Jimmy O'Bryant.) She used harmony and polyphony back and forth, one breaking away into the other, for a fine variety of textures. And she herself showed a unique ability to function as an entire rhythm section of the most appropriately inspirational sort.

On the second Austin recording session represented here, she did not top herself simply by adding W. E. Burton's drums to the group, but she did make more very good music, and I am particularly fond of the chant-like, voodoo-titled *Mojo Blues.*

Play That Thing by Ollie Powers's Harmony Syncopators is the kind of recorded performance that jazz scholars might have been picking over for years, but somehow (luckily, perhaps) they have not. For one thing it preserves some very early works by clarinetist Jimmy Noone, a rare enough event in itself for this important and influential and under-recorded musician. It also preserves Jimmy Noone playing slow blues, an event rarer still.

For another thing, the final chorus of *Play That Thing* juxtaposes two of those remarkable, traditional blues phrases that seem always to have been present in Afro-American musical lore. They are the one usually called "snag 'em" (the same one King Oliver used, in slightly different form, for his final *Snag It* chorus) and the one best known in Richard M. Jones's *Jazzin' Babies Blues* (the melody of which, contrary to popular myth, has *nothing* to do with the New Orleans Rhythm Kings' *Tin Roof Blues*).

Then there is the growth of *Play That Thing* from one take to the next. Here, for example, take 3 is introduced by the ensemble and the piano, but the slower take 4 has a saxophone introduction. Then there is the way Ladnier uses double-time: it is present in both of his solos, but is quite differently used in each. Then there is the way Noone uses some of the same ideas in both *his* solos, but to somewhat different effect.

Yet more important than any of this is the depth of the anguish in Ladnier's choruses. Let no man who hears them ever call jazz only a happy music. Let no man who hears them ever call jazz a light music. And let no man who hears them ever doubt that jazz is an important music, and a player's art. *(1968)*

King Oliver in New York*

That is how I learned to distinguish the differences between Buddy Bolden, King Oliver, and Bunk Johnson. . . . To me Joe Oliver's tone was as good as Bunk's. And he had such range and such wonderful creations in his soul! He created some of the most famous phrases you hear today and trends to work from.

—LOUIS ARMSTRONG (in *Satchmo*)

Joseph Oliver was "King" in New Orleans, not only because trombonist Kid Ory gave him the title as a billing but also because musically he *was* king. Later, in Chicago, even before Armstrong joined him, Oliver's Creole Jazz Band was celebrated for its leader's style, its excellent ensemble work, and its deep feeling. In the late twenties, Oliver, as leader of the Savannah Syncopators, had a couple of hit recordings. But by 1938—a time when trumpeters and arrangers for the big swing bands were all using his ideas—death found Oliver virtually penniless and forgotten, sweeping out a Savannah pool hall for eating money and still hoping somehow for a comeback.

A decade before his sad demise, King Oliver had been in New York where he had no permanent orchestra and little success. His followers and imitators had been there before him. The styles of his two early groups were out of fashion, and he had undertaken the then-developing "big band" approach. He was a man old before his time; he was plagued by gum and tooth troubles; he might play shakily one day and not at all the next. He was, the story goes, a shadow of his former self, using other men to play his solos and direct

* *"King Oliver in New York,"* RCA Vintage LPV-529. Reprinted with permission.

his groups, forced to record only commercial ditties, never able to recapture his former glories as a trumpeter, a composer, or a band leader.

But the story is not true, or at least it is not all true. The recordings which Oliver made for Victor between January 1929 and September 1930 have been unduly neglected. They contain very good writing and arranging, frequently good ensemble playing, and some very revealing trumpet solos by Oliver. A handful of the recognized jazz leaders did comparable work at this transitional period—the "tired old man" with his pickup groups recorded some exceptional music and played some moving horn.

We begin with the three sides produced at a very fruitful recording date. *Too Late* is a robust, double-sixteen-bar blues, with two shouting solos by Oliver on open horn. And Oliver is also probably in the lead on the final ensemble, riding over the very effective riffs. There is also plenty of trumpet on *Sweet Like This,* one of the most celebrated of the Victor series, which combines two charming themes, a 12-bar blues and a 16-bar blues. The usual opinion is that Dave Nelson plays the simple, first trumpet solo on open horn and Oliver the fine, muted variation, but it seems to me that, noting the embouchure, it may possibly be the other way around. In either case, the listener is made fully acquainted with the generous pride with which King Oliver offered his music. The other piece from this date, *What You Want Me to Do,* is an instrumental ballad on which Oliver is the only featured trumpeter. He offers a particularly touching obbligato to the theme statement of Clinton Walker's tuba.

I'm Lonesome, Sweetheart begins inauspiciously with another ensemble that might be worthy of Guy Lombardo (or, for that matter, a number of jazz orchestras at this period, including Ellington's!). It moves through a vocal by Dave

Nelson, and then we come to the first of two Oliver open-horn solos, which are the reasons for its inclusion. Here are the seminal and durable Oliver ideas of which Louis Armstrong speaks, ideas which so obviously inspired the younger trumpeter and, through him, all jazzmen.

This version of *Frankie and Johnny* is the first of two rather different arrangements of the song which Oliver did for Victor. According to guitarist Roy Smeck, who was guest soloist here and remembers the date fondly, this particular scoring proved difficult for the players, and the later arrangement was simpler. Nevertheless, the performance had a good, rocking groove at a slow tempo with Oliver solos on both muted and open horn. *New Orleans Shout,* from the same date, is one of those deliberately simple and "primitive" pieces, very much like Jelly Roll Morton's *Jungle Blues* (or perhaps like some of the recent "vamp" pieces in new-thing jazz). Once again, Oliver is the trumpet soloist.

Oliver is not a soloist on this interesting arrangement of *St. James Infirmary,* however, for here is featured the much Oliver-inspired Bubber Miley. And it was through Miley, of course, that the plunger-muted, "wa-wa" trumpet influenced Ellington so deeply.

Rhythm Club Stomp was the issued title of an interesting piece listed in the Victor files as *Curwiship Glide.* Oliver, making his statements with bravura, is the trumpet soloist (or is he playing cornet here?), and Bobby Holmes's clarinet solo and the final ensemble generate good swing. (Incidentally, some of the ideas in the writing here also show up on the Oliver-Nelson *Boogie Woogie,* a piece not in the piano blues style of the same name, and they later found their way into a pop tune about an elderly gentleman with a Biblical name who kicked the bucket, i.e., *Old Man Mose.*)

Oliver has the broad, opening solo on *Edna,* a piece in a style that might be called very good *ur*-swing band (and keep an ear out also for Clinton Walker's tuba). Since there seems to be disagreement about who plays what trumpet on *Mule Face Blues* (basically a double-16-bar piece), I'll give my own opinion that Oliver plays the first two trumpet statements, each followed by brief replies from Walter Wheeler's tenor. The long trumpet solo that follows is Red Allen in excellent form (notice his sure rhythmic ease for one thing). The brief trumpet statement at the end is probably Dave Nelson.

The next three selections again come from a single record date. It is Oliver's horn throughout the rocking *Struggle Buggy* (Glyn Paque is the alto soloist), and he is also the only trumpet soloist on *Don't You Think I Love You? Olga* opens with magnificent swing in a conversation between Clinton Walker's tuba and the ensemble. That rhythmic promise is not kept during the rest of the performance, but again we have two Oliver solos—one open, one muted.

Shake It and Break It had to be included for the way it catches a bit of the joyous side of the stomps of an earlier day. My own opinion is that Oliver does not play the "wa-wa" solos on it or on *Stingaree Blues* from the same date. These seem to me forced, slightly overblown, and limited in range in the manner of a man undertaking the style who is not really used to it. However, Hilton Jefferson plays a couple of interesting and, at the time, influential sax solos on these pieces; and Red Allen, the other trumpet soloist, is excellent on *Stingaree Blues,* a gliding, dancing statement by the last great New Orleans brass stylist—indebted to Oliver and to Armstrong, but already in 1930 a very personal soloist.

Nelson Stomp, offered in a previously unissued take, is one of the most cohesive arrangements here. It features three trumpet solos, the first and last of which are surely Oliver, but the middle (coming between Jimmy Archey's loyal parade-horn and Hilton Jefferson's alto sax) may just possibly not be he. In his solos, Oliver basically interprets Dave Nelson's melody, embellishing it with ideas typical of his work, and once again playing with a complex of pride, pain, and fine joy.

My first consideration in making the selections here was Oliver's trumpet work, and my second was the composing and arranging from the Oliver repertory of this period. Then, good solo work by others, trumpeters particularly. Probably, therefore, I should note some omissions. Oliver's Victor version of one of his great pieces, *West End Blues*, belongs to Louis Metcalfe, with a deep bow to Louis Armstrong's version of the piece. And *Freakish Light Blues, Call of the Freaks* and *Trumpet's Prayer* actually belong to Luis Russell as composer and leader, and to Metcalfe as soloist.

Here, then, is King Oliver in New York, with over a third of the recordings he made for Victor, compiled with the hope that they may be more often played, more highly regarded, and better loved than they have been. They deserve it. *(1966)*

Louis Armstrong and Earl Hines, 1928*

(For the Smithsonian album which collects the 1928 recordings by Louis Armstrong and Earl Hines, the detailed notes on the performances themselves were written by J. R. Taylor. I contributed these introductory comments.)

These thirty-two performances featuring trumpeter Louis Armstrong and pianist Earl Hines are the recorded result of one of the most important, fruitful, and influential collaborations in American music. Indeed, it is not enough to

* *"Louis Armstrong and Earl Hines,"* Smithsonian R 002. Copyright © 1975 by Smithsonian Institution.

say that their effect reached every corner of our musical culture (of whatever supposed genre), for their influence was worldwide.

In Earl Hines, Louis Armstrong had found not only a suitable companion but a co-contributor to the new music that he made out of the American past and out of his own genius. In Zutty Singleton, these two had an almost ideal drummer to complement their fresh rhythmic-melodic discoveries.

In Jimmy Noone, in his all-too-infrequent appearances here, both men had, momentarily, a clarinetist of exceptional musicianship and responsive rapport. One is grateful for the further recordings of Hines with Jimmy Noone's Apex Club Orchestra, but one may plead vainly that the clarinetist might have been Armstrong's constant companion in the studios throughout the period.

And whereas it would be foolhardy to claim that Don Redman was onto the rhythmic innovations of the young masters here, it would be equally foolhardy to deny the contribution of his steady, schooled musicianship to the proceedings in his composing and arranging—and even more foolhardy perhaps not to cite the major stylistic contributions to the emerging style of the big "swing" band.

Our collection is nearly complete, and we have favored chronology. The reader will note from the discographical data above the almost grueling nature of some of these recording sessions, as piano solos spun off from vocal dates, and small groups formed out of large—several times with the players re-using the same harmonic material to produce new pieces, and still new pieces.

We have necessarily omitted the earlier recordings with Hines and Armstrong as sidemen in Johnny Dodd's Black Bottom Stompers. And we have omitted two of the vocal accompaniments to Lillie Delk Christian. If the reader is surprised to learn that one of those is a version of *I Can't Give You Anything But Love,* he should be reassured that it adds little to the artistry (or our understanding of it) of Armstrong's other recorded versions of that piece. But *Sweethearts on Parade* here is at least a sketch for a masterpiece, Armstrong's 1930 version, and actually more than a sketch.

And, to the everlasting credit of their audiences, these recordings *sold!* Thus they must have expressed a new life aesthetic, a renewed being and way of looking at themselves and their lives, for a generation of Afro-Americans. And not only for them, but for thousands of others. For by the time these ideas had spread to other musicians, to jazzmen in general, to swing bands by the hundreds, yes, also to country-and-western singers and groups, by the time they had filtered through hundreds of pop songs and Broadway and Hollywood show tunes—by the time Armstrong's vibrato had (unconsciously) entered the vocabulary of at least every other brassman in our symphony orchestras, and by the time he had reassessed the technical resources of the trumpet and even its very function as a carrier of melody—by that time, this black man from New Orleans, Louisiana, U.S.A., who had by then also touched the feelings, and the thinking too, of millions, had altered the musical culture and the musical sensibilities of the world. (*1975*)

The Vintage Henry "Red" Allen, Jr.*

By the early '30s, Henry James "Red" Allen, Jr., had established himself as one of the most interesting trumpet soloists in jazz. He was a new addition to the hierarchy of New

* *"The Vintage Henry 'Red' Allen, Jr.,"* RCA Vintage LPV-556. Reprinted with permission.

Orleans cornet-trumpet men which included Freddie Keppard, King Oliver, and Louis Armstrong. He became perhaps the most original of all the early Armstrong-influenced stylists. He went on to become a kind of link, in style if not in influence, between Armstrong and Roy Eldridge. But Allen himself kept growing and developing as a performer and improvising soloist, and by the mid-'50s he was a greater, more flexible, more resourceful musician than he had ever been.

The preceding is, of course, part fact and part opinion. The factual part doesn't need defense. In defense of the opinions involved there is the LP inside this jacket.

Red Allen was brought to New York in 1929 by the Victor Company's Loren Watson. (There had been an earlier trip north, in 1927, to join King Oliver, but at that time the homesick young trumpeter had returned to New Orleans.) The result was a series of records, nine of which are presented here. Outside the studios Allen joined the Luis Russell band—and inside them, too, for it is actually the Russell group that passes here as Henry Allen, Jr.'s New York Orchestra. (That singular ensemble also passed, on records and on the bandstand, as King Oliver's Orchestra, Louis Armstrong's, Jelly Roll Morton's, Wilton Crawley's, and, of course, Luis Russell's Orchestra—but that is another story for another LP.)

Watson's idea had been to provide Victor with an "answer" to Louis Armstrong, and that is clearly the premise on which these sessions open. Before they are over, however, something else has begun to emerge. For his part, Allen didn't get to know this Armstrong in person down home. Speaking of his own generation, Allen says, "We got Louis from records, like all the other jazz musicians in the country, I suppose. Except, of course, we knew where his music came from better than others."

It Should Be You sketches the where-it-came-from, as Allen's lead plus Albert Nicholas's countermelody provide a modified New Orleans ensemble style. Then there are solos by trombonist J. C. Higginbotham, Charlie Holmes, Allen again, a break from Higgy, Nicholas, a break from Allen, leading into the final ensemble with more trumpet breaks. *It Should Be You* is (along with *Biffly Blues, Feeling Drowsy* and *Hocus Pocus*) one of the titles Allen himself suggested for inclusion in this set before his death.

The somber *Biffly* is introduced by chimes, probably played by Russell. The piece might be described as basically a blues in mood, but a blues built on a 32-bar outline (another example of this sort of piece is the Morton-Armstrong *Wild Man Blues*). The title comes from Allen's youthful enthusiasm for baseball; his family nicknamed him "Biff Bam," then "Biffly Bam" for his batting prowess.

Feeling Drowsy, one of the best known of Allen's early records, presents his expansive, brooding solo in a previously unissued take. There is also a frequently admired Charlie Holmes solo.

Swing Out is a fine example of the spirit of the Russell band. "I think it was the most fiery band I ever heard," Allen once said. "There was more group spirit, more friendliness, and star temperament was never a problem. It had the finest rhythm section, with Pops Foster on bass and Paul Barbarin on drums, especially inspiring to the soloists, along with Will Johnson's guitar and Russell's piano. Why, the rhythm section used to rehearse alone!" Notice Foster's response to Allen's first solo here, the bridge in Teddy Hill's chorus, and also notice Johnson behind Higginbotham.

Dancing Dave was obviously not given to the faster steps, but Allen is in fine form, showing how much of the declamatory side of Armstrong he had absorbed, and also to what understated and almost modest use he could put it.

The case for including *Roamin'* includes Allen's opening ballad paraphrase and his early ballad vocal.

With the dexterous, juggled opening statement on *Singing Pretty Songs* we are into an identifiable Red Allen trumpet style. And again the band's ensemble plus its drive and spirit are much in evidence.

Patrol Wagon Blues is a strictly dramatic utterance for the well-brought-up, virtually middle-class Henry Allen, Jr., but his heart makes it a more-than-believable one. And a personal vocal style, along with the rolling legato horn phrasing, is clearly emerging.

I Fell in Love with You features the third aspect of Allen's vocals—wordless scat singing. In the stop-time solo that climaxes the performance Allen emphasizes not a building up of individual statements, as Armstrong would have done, but an overall flow of the episode.

Stingaree Blues is a recorded appearance by Allen with his mentor, King Oliver. Allen has the final trumpet statement on open horn. And, clearly, by this time he was a quite personal soloist.

Hocus Pocus (along with other Fletcher Henderson titles such as *Down South Camp Meeting* and *Wrappin' It Up*) represents Allen at probably his most influential. Here he is suspended above the mere outline of a swing-band arrangement, with no "theme" to fall back on, inventing his own melody in his own style. Notice, also, Coleman Hawkins's lovely solo in a style that sounds quite Southwestern, à la the Herschel Evans or Buddy Tate of a few years later. (And there is splendid Hawkins still to come on this LP.)

The Crawl is chosen to represent the little blues-and-jump-oriented band that Allen (then on his own after stays with Armstrong and Benny Goodman) led during the war years, notably in Chicago. The choice of sides by this strangely neglected group was not easy, but *The Crawl* was chosen for its fine ensemble swing and Allen's typically genial good spirits, as well as its solos.

The final Allen solos here—a blues, *Let Me Miss You, Baby,* a slow ballad, *I Cover the Waterfront,* and a medium-fast standard, *Love Is Just Around the Corner*—come from 1957 and they represent this exceptional musician at full development. Dynamics are effective child's play to him: he moves from a whisper to a shout in a single phrase with complete aesthetic logic. Long, asymmetric phrases roll by with ease. His rhythm and swing are just about impeccable. And his ability to set and sustain a mood is exceptional—all the more so when one realizes the adventurousness, the emotional variety, and the diversity of tone color and trumpet sound an Allen solo is apt to encompass.

The horn techniques involved in this sort of playing attracted musicians of all ages and persuasions, and they soon had young trumpeter Don Ellis calling Allen the most avant-garde of all jazz trumpeters. But the feeling that dominates his music he once explained this way: "Playing is like somebody making your lip speak, making it say things he thinks . . . it's a home language, like two friends talking. It's the language everybody understands." (*1968*)

Dizzy Gillespie: The Development of an American Artist*

(The notes for the Smithsonian's two-LP collection of Dizzy Gillespie's early works include copious quotations from Gillespie himself from a variety of sources. Here I have either deleted those quotations or, when it seemed advisable, indicated their contents in paraphrase.)

• I •

This album collects recordings done over a crucial five years by an American trumpeter and sometime composer-orchestrator.

What emerges in the most unassuming of ways—and, for some of us, the most unexpected of places—is a pattern of musical discovery. Also, if you will, self-discovery. But the special self-discovery of an artist, hence of someone who makes discoveries for us all about our selves, our lives, and our possibilities.

We begin with Dizzy Gillespie emulating his idol and guide, Roy Eldridge. We hear him become himself as a master soloist, leader of small ensembles, composer-orchestrator for big bands. We hear him discovering a new jazz idiom and learning to dramatize it, not only for himself but within ensembles of various kinds.

We have carefully avoided here all the recorded collaborations with Charlie Parker. They have often been collected and reissued, and, more important, Parker's brilliance has sometimes clouded the issue of Gillespie's own.

* *"Dizzy Gillespie: The Development of an American Artist,"* Smithsonian R 004. Copyright © 1976 by Smithsonian Institution.

• II •

In the mid-1970s, Dizzy Gillespie sees himself—impersonally, almost reverently—as a part of a brass tradition which began with King Oliver's cornet and moved through Louis Armstrong, Roy Eldridge, and himself. He sees that heritage as also having been passed on to others.

The United States is 200 years old and for over half that time there has been a clearly identifiable Afro-American music. Almost six decades ago there began to emerge the first evidence of a "classic" tradition within that body of music, the musical idiom that came to be called "jazz." And beginning in the 1920s, jazz recordings have documented the music of some of our greatest instrumentalists and best composers.

It seems quite safe to say that Afro-American music in general and jazz in particular have changed the musical sensibilities of the world for musicians and listeners alike.

But, aside from the body of recorded music it has offered, jazz has also contributed to instrumental discovery and instrumental technique for composers and musicians of all kinds in all idioms, particularly for the brass instruments, and most particularly for the trumpet.

In some respects that contribution is obvious. In range, for instance, a workable two octaves above middle C is now virtually commonplace, with some trumpeters able to reach even higher with precision. Some jazz cornetists and trumpeters have also gone lower than the previously accepted limits on the horn. And they have worked out half-valve, "choked" effects as well. To the single European "straight" mute, American players have added a variety of other mutes and combinations of mutes, played both in place and variously manipulated over, around, and in the bell of the instrument. (That latter does not apply to Gillespie however, who prefers to play with open horn or occasionally with a cup mute.)

But more important perhaps is a change in instrumental function for the trumpet. The point is subtler and more elusive, but I think that through the sensibilities of jazz musicians, the trumpet has become a carrier of melody—flexible, lyric, or virtuoso melody—in a way that it was not before. Laying aside questions of the way notes are articulated in jazz, it seems to me that brass melody of the quality one hears in this collection on *Star Dust I* or *I Can't Get Started* or *Be-Bop* would be hard to imagine within a strictly European musical tradition.

All of which means that John Birks "Dizzy" Gillespie is a great figure in American music, in world music, and perhaps the greatest living musical innovator we have.

• III •

In the recorded selections which follow, once we have left the well-established idiom of Cab Calloway's big band working for a major label, the recorded evidence of Gillespie's career and growth is sometimes a matter of luck, of happenstance, of dedication, of idealism (sometimes misguided idealism), of opportunism. And the recordings sometimes reflect the realities of the nightly bandstand and sometimes do not.

It is one of the great pieces of luck of American musical history, surely, that a young Jerry Newman carried his portable disc recorders into Harlem clubs in 1940 and 1941 and preserved real jam session material by major figures. In the months between the selection by Lucky Millinder's big band, *Little John Special,* and the Coleman Hawkins session that produced *Disorder at the Border* and *Woody 'n' You,* the American Federation of Musicians had banned all recording activity by its membership unless the companies signed certain payment and royalty agreements. The major companies held out, but small labels willing to sign with the A. F. of M. sprang up right and left, and (behold!) they recorded jazz, and quite often the new jazz of the 1940s.

Did they know they were preserving for posterity portions of a great period in American music and the emergence of a kind of musical miracle? Surely if they did, they would never have thought to put it that way. So perhaps a secondary dedication here should be made to the proprietors of those record companies for their taste, their idealism and their enthusiasm, their insight—and yes, for their opportunism as well.

• IV •

Gillespie has spoken often of the value of the varied experience he had in his early professional life with Teddy Hill's Orchestra, and with Cab Calloway. In writing *Pickin' the Cabbage* for Cab Calloway in 1940, Gillespie took keen advantage both of Milt Hinton's exceptional craftsmanship (a craftsmanship far beyond that of most jazz bassists then—listen below) and of Andrew Brown's baritone saxophone, newly added to the band. On the surface of course *Pickin' the Cabbage* is a typical swing band riff-tune, but an oddly distinctive one.

Gillespie's solo is striking. Its accents and phrasing, its melodic rhythm, are very close to Eldridge. But the choice of notes, and the harmonic savvy involved, are the work of the young Gillespie. The solo is also a cohesively durable improvisation, and, granting it one slight flaw-in-passing, is very well played.

Hard Times is a vocal novelty by Calloway, and thus a part of the reason for this band's existence. It also features the group's most popular instrumentalist, tenor saxophonist Chu Berry, and a Gillespie who plays far more like his youthful model than on the previous piece.

Bye Bye Blues was the Calloway band's fast showpiece—a kind of answer, if you will, to the Basie band's *Twelfth Street Rag* or to any of a number of things in the Lunceford book—with Gillespie as a showpiece Eldridge. The vibraphone soloist is of course trombonist Tyree Glenn. And the exultant riffing that takes the piece out shows the level of this band's ensemble work.

Our excerpt from *Boo-Wah Boo-Wah* comes from another vocal novelty featuring the leader, and was in turn knocked together from a muted-trombone and/or trumpet section cliché of the period. It is included here because of the striking mélange of styles in Gillespie's solo, which starts out as Eldridge, goes to an embryonic version of what later became one of Gillespie's most characteristic and imitated phrases, and (after Tyree Glenn's trombone solo) settles almost comfortably into swing phrasing but Gillespie choice-of-notes.

That latter approach dominates the remarkable jam session material recorded in Monroe's Uptown House after hours in 1941. The two versions of *Star Dust,* one cup-muted throughout, the other on open horn, show not only a particularly thoughtful young ballad player but, in the second version, an inspired one. Indeed, the sober, introspective half-chorus at his second entrance on *Star Dust II* might stand with Gillespie's great ballad improvisations. Yet, compare him with Don Byas here. Byas is already a master player in command of his own probing, Coleman Hawkins-based style. Gillespie is only at the beginning of his musical self-knowledge.

Incidentally, those two solos on the piece titled *Kerouac* (actually improvised variations on *Exactly Like You*) are both parts of a single long performance. We polled a careless sampling of both writers and musicians on which *Kerouac* solo to include and we got two sets of answers: the writers generally favoring the more sedate first solo, the musicians the more adventurous second. We decided to settle for both. (The piece was titled by Newman, by the way, for a young fan and hanger-out whose later theories on literary style showed his envy of the spontaneity of these musicians.)

With Gillespie's half-chorus on the Les Hite band's version of *Jersey Bounce* (1942) (again, variations on *Exactly Like You*) we are thrust briefly into a different musical world. What is the nature of the difference? And why would musicians and listeners alike soon need a new catch-word—*bebop*—to name that difference? For the same reason they had needed words like *cakewalk, ragtime,* and *swing* to name earlier styles in Afro-American music.

The real difference in what Gillespie plays here, the newness of it, is rhythmic. Where did it come from?

Gillespie is quite frank about the source of it in his own musical evolution. "I learned rhythm patterns from Charlie Parker." Still, hearing the masterful way the trumpeter handles rhythm in this and subsequent solos, we are also aware that he made Parker's rhythmic language his own. And beyond that, we sense that both men were in touch with something deeper, less personal, and more universal.

This valuable (but rather badly recorded) Les Hite selection did not find Gillespie merely sitting in for a record date. The Los Angeles-based Hite band, which had earlier featured a young Lionel Hampton and also briefly served as Louis Armstrong's band, was in the East on a tour where Gillespie joined them. Clearly pianist Gerald Wiggins was a valuable man among the regular members.

Little John Special is also evidence of a regular gig for Dizzy and finds the Lucky Millinder group rushing into the studio just before the A. F. of M. recording ban took effect.

The piece is a veritable playground for a jazz historian

with pedantic urges. The title is a switch on Fletcher Henderson's *Big John Special,* which was in turn named for a well-known 1930s Harlem doorman-bouncer. But the piece is a Basie-style blues with its opening riff borrowed from *Boogie Woogie* (*I May Be Wrong*) and one of its closing riffs borrowed from the *One O'Clock Jump,* which also no doubt encouraged *Little John*'s internal modulation. But a real point of interest is the presence of the *Salt Peanuts* riff, executed, accented, pronounced so differently here than in its later version below. The opening phrase of that riff can also be heard as one of Louis Armstrong's breaks on his 1930 *Ding Dong Daddy,* but I suspect the most immediate origin of the figure *and* the octave jump response is again Basie, who used both licks during the period.

But there's more. When he heard this recording in 1973, Gillespie remarked quietly, *"Salt Peanuts* AND *Cool Breeze!"* (Hang on to that riff that accompanies his solo.)

The other soloists include Tab Smith on alto and Ernest Purce on baritone, and the clarinet solo is probably Stafford Simon.

• V •

What we have heard on the Millinder selection is a developed modernist solist somehow featured on a swing band blues. Our next selection, recorded over a year later and after the A. F. of M. ban, takes another step.

Woody 'n' You is a major Gillespie piece, and a major jazz composition in the sense that almost every young player since has known it and learned to meet its harmonic challenges.

This was the first recorded version of *Woody 'n' You,* and even with the cut-down brass here (there are no trombones) it was the first large-ensemble modern jazz recording. The title comes from the fact that the piece was originally written for Woody Herman's band. What makes it a Gillespie piece? Its chord progression, the implicit "Latin" character of its accents (even when played, as here, without a Latin beat), and perhaps the fact that (like so much of his composing) it is in a minor key.

Hawkins, like his "pupil" Byas, was a soloist whose phrasing basically belongs to the 1930s (perhaps to the very early 1930s at that) but whose harmonic knowledge and imagination allow him to meet all the challenges which the younger players were coming up with in the 1940s.

Gillespie's own leaping *Woody 'n' You* solo is still a bold statement, but now also a classic one. And much of the middle section of the blues *Disorder at the Border* was incorporated into *Cool Breeze,* with a part of what Dizzy plays solo here taken by the full ensemble.

The sessions that produced these two recordings, by the way, were tough ones. Except for the three Hawkins ballads they produced, the music was difficult for the players, difficult not so much for the notes themselves as for the new ways of phrasing those notes. The sessions were also the first recording experience of Max Roach, then a very young and (he says) very nervous musician whose subsequent work of course discovered and defined the proper role of percussion in this new music.

Billy Eckstine has explained that when he formed his band in 1944, the group had only two pieces of music. "The two scores were *Night in Tunisia* by Dizzy and a thing Jerry Valentine had written. . . . We really whipped the band together in St. Louis. . . . Tadd Dameron had moved into Kaycee at that time, and when we got *there* he used to work along with us, writing some things for the band like *Cool Breeze* and *Lady Bird.*"

But the records were made mostly out of a conviction

(perhaps the label's conviction) that some blues plus lots of Eckstine ballads were a surer commodity. So we get precious little Gillespie on these early Eckstine discs—but precious it is.

Still, the trumpeter's spirit is almost everywhere and there are some solos. Take the ascending brass responses (sympathetic, cajoling) during the leader's second vocal chorus on *I Stay in the Mood for You.* And there is the brass behind Eckstine on the "tenor sax battle" specialty, *Blowing the Blues Away.* Gene Ammons begins and Dexter Gordon plays his antagonist. The piece is taken up and then out by Gillespie, then by Al Killian—all too briefly.

John Malachi readily acknowledges *Opus X* as a four-beat adaptation from Eddie Durham's two-beat *Lunceford Special* (1939). Gillespie has the opening bridge. The alto sax solo is John Jackson and Al Killian's trumpet takes this one out. Whatever the shortcomings and frustrations of these records the spirit and drive of this transitional ensemble comes through strongly.

While he was with the Earl Hines band in 1942, Gillespie first wrote out the piece one night in Philadelphia, using the top of a trash can as a kind of makeshift desk. Indeed, it was Hines who named it *Night in Tunisia.* The label of this Sarah Vaughan vocal version of *Night in Tunisia,* called *Interlude,* originally carried the alternate title, as did the instrumental version by Boyd Raeburn heard later in this album.

Leonard Feather, who produced these Sarah Vaughan recordings, remembers that Dizzy approached him with a demo record of this *Tunisia*-with-words by the young Sarah Vaughan, who had sung both with Hines and with the Eckstine band. The fine little arrangement is of course Gillespie's. Sarah Vaughan was then only beginning to discover the vocal resources which today make her one of the great singers.

This rare "take" of *No Smoke Blues* (Feather's blues on the wartime cigarette shortage) is delightfully looser and freer than the original on the part of both singer and trumpeter. Also it does not make use of the riff later called *Blue 'n' Boogie,* as the more familiar version did.

Finally, don't miss the way that Byas opens his solo by mimicking Dizzy's final lick.

• VI •

The recorded evidence of the heightened musical activity along "the Street," Fifty-second Street off Sixth Avenue in New York, in 1944 dates from after-the-fact and has an oddly altered personnel. But it is remarkable nevertheless.

We begin with an Oscar Pettiford ensemble, a large studio band which somehow recorded only this one instrumental. It plays with exceptional ensemble spirit and swing for such a group.

Something for You was also known as *Max Is Makin' Wax*—the Onyx quintet had used it as a vehicle for Max Roach's drums—and later also as *Chance It. Something for You* is virtually an archetypal big band bop arrangement. If it had had a better theme (it is a rather static piece until it digs into the opening bridge) it might have been an almost perfect one. Gillespie's solo begins with an exuberant shout which sets the tone for the rest of it. And notice the riff that accompanies him—the Woody Herman band picked that one up.

Pettiford's immaculate skill and artistry as a bassist resounds on this and several subsequent selections. But one should not compare him (nor compare Ray Brown below) only to the thumping and seemingly tone-deaf bass players heard in some of these groups. Remember the beautiful Mr. Hinton where we started.

With *I Can't Get Started, Good Bait, Salted Peanuts,* and *Be-Bop* from 1945, we come to the first recordings Gillespie made under his own name. *I Can't Get Started* is the earliest of his several recorded versions of that ballad (which have delighted its composer, Vernon Duke, by the way).

With Gillespie's penetrating lyricism on *I Can't Get Started,* and with his amazing break and witty exuberance on *Salted Peanuts* (a better solo than the later one on the recording of the piece with Charlie Parker, I think), and with his staggering three-chorus *tour de force* on *Be-Bop*—with these three things alone, Dizzy Gillespie might easily have established himself not only as an original soloist but as a major musician. (And *Salted Peanuts* and *Be-Bop* both out-of-print since the days of 78 r.p.m.!)

As one might expect, Byas whirls through the chord changes with his imaginative, arpeggio-based style. What is not so expected perhaps is the ease and the verge with which he plays the unison ensembles in the accents of bop. The bridge on *Be-Bop* is a glory and the ending a delight.

More *Peanuts* pedantry: the piece was slight mistitled here, no doubt by the record company. It uses an *I Got Rhythm* chord progression (a tenor group—Coleman Hawkins, Georgie Auld, and Ben Webster—had so recorded it in mid-1944). Here it has also been outfitted with a striking introduction and ending, an interlude, and a mood of sprightly burlesque and, under that, of sublime seriousness.

Gillespie did a record date in 1945 with (of all people) clarinetist Joe Marsala and what was surely one of the most incongruously assembled studio groups in recording history. Pianist Cliff Jackson came away from the date shaking his head because the trumpeter spent much of his time playing higher notes than the clarinetist-leader. Certainly the two ends of Fifty-second Street met here: Marsala led a traditional group at the conservative Hickory House, and Gillespie was fairly recently out of the Onyx Club basement, a block and a half (and two musical generations) away. We have picked two Marsala selections on which the trumpeter is featured: his bold embellishments and inventions on the rare *On the Alamo* and this very interesting but all-too-brief version of *Cherokee*. By this time Charlie Parker had virtually established his proprietary right to that latter piece through his brilliant use of its chord progression on *Ko-Ko*.

With the Boyd Raeburn band Gillespie and Pettiford were recording-date guests and received featured billing on the original record labels. *Interlude* retains the title of the vocal version of the piece but restores the original instrumental treatment. This is Gillespie's own *Night in Tunisia* arrangement as done for the Earl Hines band, with Walt Robertson taking the trombone opening originally done by Benny Green. Gillespie enters on the famous "break" that was later doubled and then tripled in length, and he offers a showpiece solo and coda despite a somewhat unswinging performance by the ensemble.

This is not the place to go into a history of the Raeburn band (or Raeburn bands, actually), but it is perhaps the place to correct a long-standing error. As Raeburn historian Jack McKinney points out, and Ed Finckel confirms, it was Finckel who wrote the blues *March of the Boyds* and not George Handy. Handy's steady association with the band was established after Raeburn moved to California. The tenor saxophone soloist, by the way, reaching into a modern style, is Joe Magro.

• VII •

On the two Georgie Auld selections, Gillespie is sitting in with two studio-assembled efforts to launch the tenor saxophonist as a band-leader. They offer two sides of the stylistic question as of the mid-1940s, both from the same composer-arranger (and guitarist) Turk van Lake (née Vanig Hovsepian). *In the Middle* is a swing-band arrangement; *Co-Pilot* is an attempt to build a big band bebop score more or less on the chord progression to *Be-Bop.* (Gillespie accomplished the latter task himself with the astonishing *Things to Come* in 1946.)

On *Co-Pilot* Erroll Garner is an original delight, and fully in context. And on *In the Middle,* notice the way Gillespie almost audibly replaces his horn in his embouchure after his first, relatively cool eight measures.

But between these two Auld selections, Gillespie had had another record date with himself as leader, and from it comes the rarest selection here, the fascinating *Groovin' High* (variations on the *Whispering* chord progression) recorded with Dexter Gordon. It is rare because it comes from the only known copy of this particular "take" of *Groovin' High.* The recording itself will explain its interest, but do not miss the way that Gordon counterpoints a portion of the opening "head" against Gillespie's statement. What is left to add is that immediately after M. D. Zemanek taped his worn 78-r.p.m. copy of this *Groovin' High* for inclusion here, the walls of one of its grooves broke down forever.

Our final selections come from Gillespie's 1945 Los Angeles trip, an effort of Billy Berg's club on the Sunset Strip to offer Californians the new jazz that had begun attracting attention in New York. The personnel more or less speaks for itself, with a young Milt Jackson, already a formidable soloist, but, as one wag put it, sounding like he is playing on a bunch of milk bottles compared with the way he sounds today. Lucky Thompson was hired in L.A. as a kind of standby to Charlie Parker, who was unable to show up for the job almost as often as he was able to.

The records established the Dial label for Ross Russell,

then proprietor of the Tempo Music shop in Hollywood, and were made pseudonymously because of contract difficulties as the Tempo Jazzmen. Gillespie was originally billed on the label as "Gabriel." He complained during the date, by the way, that the recording studio had such a low ceiling that he couldn't turn the bell of his horn up and hear himself properly.

Introduced by the riff which was later functionally known as *Oop-Bop-Sh'-Bam,* this *Confirmation* is the major version of Charlie Parker's most brilliant piece (and his only piece with an original chord progression). Gillespie's keenly paced solo does worthy credit to the work and its maker.

The searing *Dynamo A* is a retitled *Dizzy Atmosphere* as indicated, and (some say) actually Thelonious Monk's line and a kind of exchange for Dizzy's having contributed the "double-diminished" verse to Monk's *'Round Midnight.*

Gillespie handles his music now like a master. He coolly, tastefully approaches his fiery bridge on *Confirmation.* He handles the changes of Richard Rodgers's *Lover* on *Diggin' Diz* with wit and melodic ease. He ponders *'Round Midnight* with creative introspection. He is now the master musician ready for high public success and world fame.

What we have witnessed here is a remarkably well-documented development. But lest we think otherwise, still only a beginning. There was, and is, more to come in instrumental mastery, rhythmic flexibility and ease, range of dynamics, ideas, and feeling, both earthy and refined. "I was developing. I am still developing," Gillespie has said. "He was dizzy like a fox," said Teddy Hill. (*1976*)

Miles Davis: Odyssey!*

The music on this LP first appeared as *"Miles Davis and Milt Jackson: Changes,"* and the record was made at an auspicious time for several of its participants. Miles Davis was shortly to form and record with the Quintet with John Coltrane, the group with which he first achieved national fame and celebrity. About a year before, he had recorded *Walkin'* and *Blue 'n' Boogie,* and, following them, *Bags' Groove* and *The Man I Love,* records which led both musicians and the more hip and "inside" fans to a virtual rediscovery of Miles Davis.

At the same time, Milt Jackson and Percy Heath were reasonably fresh from the first real successes of the Modern Jazz Quartet, and from such personal triumphs for Jackson as his work on *Django, Autumn in New York,* and (going back a bit earlier) *All the Things You Are* and *Delaunay's Dilemma.* Also, Jackson and Heath had both been participants in the aforementioned *Bags' Groove–The Man I Love* "all star" date with Davis.

So, on the basis of the moment it was recorded in the careers of these men and the development of their styles, I think this music should be better known and more widely discussed than, to my knowledge, it has been. At the same time, I find this an intrinsically delightful session on its own terms, and that is at least an equally good reason for my thinking it should be better known and more celebrated.

"Changes" are of course chord changes to a jazz musician, and hence they are a pattern, an outline, a design on which he is to improvise. But without doing too much damage to the meaning of the word, we might say that changes lead to differences. And the differences between Bryant's lovely and unusual piece, *Changes,* which closes the program, and the

* *"Miles Davis: Odyssey!,"* Prestige 7540. Courtesy of Milestone and Prestige Records.

Jackie McLean piece, *Dr. Jackle,* which opens the LP, seem to me wholly remarkable. Yet both pieces are in the same form, since both are basically the 12-bar blues.

McLean's *Dr. Jackle,* the faster of the two, has a title which puns on the nickname and professional name of the alto saxophonist born John Lenwood McLean and on Robert Louis Stevenson's character. Subsequent errors have wrongly titled the piece both *Dr. Jackie* and *Dr. Jekyll*—that second title is the one it was last stuck with. The piece long remained in Davis's repertoire, and for all I know he still uses it. But later versions have turned it into a breakneck showpiece of quite a different mood than can be heard in this performance. Davis's energy, particularly, has a fine air of gaiety and playfulness—even joyfulness—without being either glibly or superficially joyful. The theme itself is interesting: it opens with the kind of leaping intervals that forecast some of McLean's more recent work, but it settles into a more conventional groove, reflecting the music of McLean's youthful idol, Charlie Parker, plus a touch of Miles Davis.

Ray Bryant's slower *Changes* has no theme, but is a series of finely patterned, substitute chord changes on a blues outline. The effect of Bryant's skeletal chords plus the talents of the individual players is to evoke an almost ballad-like, lyric mood. That mood is perhaps unique in blues recording, but in any case, the performance is never out of touch with the fundamental strength and unsentimental emotion that good blues ought to have. It makes a striking vehicle for Milt Jackson, who has proven himself a great blues player time and again—he does on *Dr. Jackle*—and who is also a great ballad player. Here on *Changes* the two abilities justly come together.

Note, therefore, that in these two pieces the players have discovered, within the blues form, a wide range of expression. But not once have they resorted to the sort of rustic, "just folks" posturing that a great deal of self-consciously funky "soul" blues has offered.

And these remarks, I must admit, lead one to direct praise of Ray Bryant's work on these tracks. He had recently arrived from Philadelphia when they were done. There he had been "house pianist" at the Blue Note, working with men like Davis and Charlie Parker. And now he was in New York associated with some of the best jazz men. His work on this LP shows that he fully deserved the compliments such associations imply.

Bryant was still in his early twenties when this session was done, as was Jackie McLean. But McLean was well known to most New York modern jazzmen when he was still a teenaged schoolboy. He played with a neighborhood group which included the likes of Sonny Rollins and Kenny Drew; he studied after school with Bud Powell. He was a Charlie Mingus sideman during the year of these selections, but he had made his first recording with Miles Davis in 1951, *"Diggin' with the Miles Davis Sextet."* McLean's second contribution as composer and player here is *Minor March.* The piece is more Bird-like, even to the themeless, improvised bridge, a practice Parker was fond of. Jackson takes that bridge in the opening chorus, and he, particularly among the soloists, does a lot of interesting things with the bridge in his choruses. And Davis's solo builds from an almost wispy beginning through some fine, understated staccato exuberance.

In the original notes to this music, Ira Gitler called Thad Jones's piece, *Bitty Ditty,* a "piquant" line. Its understructure generates an unusual feeling of suspense in the solos. On the face of it the effect would seem particularly well suited to Davis, and he solos twice on this one. Notice also how much real swing there is in Bryant's left hand alone during his choruses, and how Jackson employs one basic rhythmic idea during his.

A final word about Miles Davis's solos on these pieces. They are marvels of the kind of contradiction of which he was so eminently capable at this period. He seems to play in relatively short, asymmetrical musical "breaths"—to pile up phrases, each upon the other. Yet there is seldom any doubt about the logical linkage of those phrases, nor about the continuity and design of his solos as wholes. His style may seem simple, and it certainly seems to be executed with great ease and fluency, yet it is full of far-from-simple and highly personal effects. And the sound Davis evokes from his horn is often the result of the most delicate, skillful, and interrelated manipulation of his lip, perhaps a mute, his position at the microphone, plus—let us admit it—the innate and unaccountable musical personality of Miles Davis.

The years 1954–58 were important ones in Davis's music. This LP is further evidence of how important. *(1967)*

Charlie Parker: The Dial Recordings, Volume 2

(These notes were written for a proposed second collection of the records Charlie Parker made for the Dial label in the mid-1940s, specifically centered on the March 28, 1946, recording date with Miles Davis, pianist Dodo Marmorosa, tenor saxophonist Lucky Thompson, and others. The LP was never released, but the music in question has been issued on Spotlite, Warner Brothers, and other labels.)

This LP begins with the first recording done by Charlie Parker when he arrived, as a member of Dizzy Gillespie's group, for a turbulent engagement at Billy Berg's in Hollywood. The engagement was received with enthusiasm by musicians and certain younger fans, criticized by others, and overtly attacked in the press. A hard time for Gillespie's group and a harder one for Parker, who stayed on after they had left. The LP also offers the brilliant first date Parker made for the Dial label on his own. It then bridges the time Parker spent at Camarillo State Hospital after a breakdown, and concludes with the first recording he did after his release, Parker's solos from a jam session taken down on a home machine.

The set also offers performances of one of Parker's most effective compositions, *Moose the Mooche.* And it has, in the fourth take of *Ornithology,* one of the generally acknowledged masterpieces of the altoist's recorded career.

Diggin' Diz comes from a rehearsal. It was a rather chaotic occasion held in a little studio attached to a church in Glendale, California, which was jammed with milling fans and hippies—the word had got out that Dizzy and Bird were to play. Basically, the group was Gillespie's, the one that he had taken into Berg's, with the exception that George Handy, who was supposed to contract the record date, substituted on piano (rather unswingingly, I'd say) for Al Haig, and Milt Jackson didn't participate. The performance is, understandably, rather disorganized, except for Gillespie's solo, but there are few enough recordings of Parker and Gillespie from the period when they played together regularly, and any one of them is more than worth preserving. When the date itself came off, a few days later on February 7 to be exact, Parker didn't show up, and producer Ross Russell asked Gillespie to take charge himself, which he did. *Diggin' Diz,* sometimes called *Diggin' for Diz,* adopts the chord structures of a rather unexpected Broadway waltz, one which was rather widely played by young jazzmen at this period, Richard Rogers's *Lover.*

Ornithology, and the rest, come from the first official Parker date for Dial records. Bird was especially enthusiastic about the availability for these recordings of trumpeter Miles Davis, who came through Los Angeles with Billy Eckstine's band. *Moose the Mooche* is a rhythmically ingenious melody. *Yardbird Suite,* based on the chord progression of Earl Hines's *Rosetta,* was, according to reports, written by Parker for the Jay McShann band years before. You will notice that for this one, only the first and fourth takes are preserved; possibly the second and third were too quickly interrupted.

Parker somehow does not solo on the first take of *Ornithology,* but the group finished it out, perhaps to rehearse it, with Dodo Marmarosa taking Parker's space. This version is worth preserving, partly for completeness, and partly also for evidence that, as on the majority of these performances, the other players often use variants of similar solos on each take of each piece, whereas Parker does not. The first take also exposes Marmarosa's ideas for the piece—and these are interesting because of the not-always-ideal support of, say, Bird's implied chords on the later takes. But of course Parker's ideas were then difficult for most other players in jazz. The theme here is an elaboration, by trumpeter Benny Harris, of a phrase Parker recorded on *The Jumpin' Blues* with Jay McShann. Harris developed it on the chords of a well-known modernist's standard, *How High the Moon.* One might also note that Lester Young had played that same opening phrase in 1936 on *Shoe Shine Boy.*

The most difficult piece was apparently *A Night in Tunisia.* It took perhaps two to three hours to record, according to Russell. Parker's break and subsequent solo on the first take were so brilliant that they were issued as a fragment, but the rest of the performance was too ragged to use, and it was not until the fourth attempt, you will notice, that a usable take was made.

At the end of this date, there is a gap in the Parker Dial series, which we do not propose to fill in this series of LPs. It was the famous *Lover Man* date, at which Parker was in very bad physical and mental shape, and at which he made a desperate, agonizing effort to play well, and after which he suffered the breakdown that sent him to Camarillo. The records have appeared, to be sure. They will not appear here. And let that be the end of it.

We conclude with some solos by Parker, made almost immediately after his release from the hospital, and before he had done any studio recordings. These come from a session held at the home of Charles Copely. Also present, incidentally, were trumpeters Howard McGhee, Shorty Rogers, and Mel Broiles. The latter is now lead trumpeter with the Metropolitan Opera Orchestra, and he searched his memory for the particulars of these home recordings. They have been issued before, the first two with *Home Cooking* titles, the third with no title. (It is *possible,* by the way that this third solo was not recorded during the session at Copely's, but comes from a later session.)

We should be a bit less peripheral in our comments about certain of these performances, to be sure. I was particularly struck by the fact that Parker's time and phrasing are so impeccably sure on some of the early takes, before the balance is set and before the group as a whole, or the rhythm section as a part of it, has the piece in hand. The playing of drummer Roy Porter needs comment, I think, a somewhat negative comment unfortunately. At the time Porter was drummer with Howard McGhee's group, and one of the few in California to court the then-new jazz idiom. He understands that the drummer's role involves spontaneous counter-motifs and accents which comment on the improvisation. But Porter does not always seem to be a sensitive compliment to the players; his bass drum "bombs" and snare accents often seem to me to break up the flow of the music. However, his eagerness and energy are otherwise an asset to be sure.

Moose the Mooche is not only one of Parker's most highly praised lines but it has fathered several other compositions, including Sonny Rollins's *Oleo,* I would say. Critic André Hodeir has particularly praised Parker's harmonic ingenuity here. He was speaking of the bridge of Parker's solo in the second take which achieves an effect of polytonality—that is, a suspended effect of being simultaneously in more than one key. Hodeir comments in *Jazz: Its Evolution and Essence:* ". . . Parker grafts a major chord based on the sixth degree of the scale onto a dominant seventh, thus forming an altered thirteenth that suggests two different keys, even though the notes played do not violate the laws of natural resonance. This passage (the bridge) may be profitably compared with the corresponding point in the following chorus, played by Lucky Thompson, who is an excellent musician but who is unable to follow Parker's lead in shaking off the bass's harmonic tyranny."

The more tender Parker of the *Yardbird Suite,* of both the theme and the alto solo is, understandably, the favorite Parker of Lee Konitz. I am particularly intrigued by the slight but appropriately lovely differences in the way Parker phrases the bridge in the first chorus, which he plays solo in both takes, in slight paraphrase of the line as written. Perhaps this date on the whole is most interesting because one discovers that Parker's coolness on *Yardbird* was not a matter of his approach for that day, or his mood of the moment these takes were made, but of what he thought appropriate to that piece. For there is a very different Parker on *Moose,* on *Ornithology,* and on *A Night in Tunisia.* And those who will not allow that Bird had that kind of artistic discipline should listen carefully.

When this fourth take of *Ornithology* was first issued,

it almost immediately became a legend among musicians. Never before (or since?) had Parker's ingenious rhythmic originality been so dramatically evident. For one thing his phrasing—rapid as it is!—is an ingenious alternation of long/short/long/short lines, with some rests in between. Perhaps more important, his accents occur on the beat, the off-beat, before the beat, and so forth. For this solo, André Hodeir has done a rather extensive analysis in his book *Toward Jazz,* and he concludes, "Rhythmically speaking, the outstanding passage is that which extends without interruption from bars 16 to 20 (neglecting, therefore, the double bar at the end of measure 16); in it a cell consisting of a displaced rhythm (measure 16) is expanded, so to speak, by the addition first of one, then of two note values (measures 17 and 18)." The final wonder is that a passage which includes such variety and fundamental asymmetry can have the striking effect of self-containment and completeness as this one.

Perhaps the daring in each take of *A Night in Tunisia* can speak for itself, but, again, the spontaneity with which Parker handles an alternation of tension and release, complexity and simplicity, might take days for a more self-conscious artist to work out on paper. In passing, notice also how the young Miles Davis echoes Dizzy Gillespie on this, Gillespie's piece.

One final remark on the post-Camarillo jam session tracks, specifically on the one titled *Home Cooking II,* which, like Parker's *Ko-Ko* is actually variations on the chord progression of Ray Noble's *Cherokee.* It is not so great a performance as *Ko-Ko* (and there is a bit of reed trouble), to be sure, but it is particularly interesting as a contrast, I think, for here is the virtuoso Parker on one of his show pieces, but in a mood of comparative serenity. (*1962*)

*The Modern Jazz Quartet: Plastic Dreams**

If this LP were nothing else (it's a lot else), it would be evidence of the enviable, rewarding durability of perhaps the most illustrious small ensemble in jazz history. Think about it: the Modern Jazz Quartet has been with us for nearly twenty years. No other small ensemble of relatively stable personnel has had such a sustained career. Indeed, no other jazz ensemble of any size except the Duke Ellington orchestra.

The reasons for the MJQ's durability are not at all obscure or difficult. They start with the fact of four individually exceptional jazz musicians, who also mesh together spontaneously in such a way that they can even anticipate each other's surprises—and take delight in the anticipation.

Prove it to yourself. Simply put your playing stylus down (carefully now) in the middle of *Walkin' Stomp* or *Dancing* or *Piazza Navona* and hear and feel the playing, the interplaying, the impeccably infectious but subtle swing. It is these qualities that attract the casual listener as well as the committed fan. It is these qualities that have made the MJQ a source of pride to so many of us—pride that jazz has an ensemble which plays the music with such ensemble perfection. And it is these qualities that have attracted musicians of all persuasions the world over to marvel at the MJQ.

Perhaps a brief history of the group is in order. It began, actually, as the rhythm section of Dizzy Gillespie's 1946–48

* "*The Modern Jazz Quartet: Plastic Dreams,*" Atlantic SD 1589.

big band. Indeed, in 1948, its members made a rather obscure record date, under Milt Jackson's name, and with the celebrated Cuban bongo player, Chano Pozo.

As John Lewis has modestly explained, it was the quality of their work on this and subsequent recording dates together (done in 1951 and 1952) that convinced the members of the group that they liked playing together and wanted to continue to.

Of course it was not that easy. The musicians had to bide their time and take other jobs before they could persuade the rest of the world of what they had already discovered for themselves about their potential. Jackson worked with Woody Herman and was back with Gillespie. John Lewis accompanied Ella Fitzgerald and worked with Charlie Parker and Lester Young.

But the recordings together continued, and, with John Lewis appointed musical director of a cooperative group, the style grew. Then, in early 1953, one of the group's LPs caught on with record buyers, and the Modern Jazz Quartet became a going thing in nightclubs and concerts.

There was one change in personnel. In early 1955, the great drummer Kenny Clarke decided to leave the United States and take permanent residence in Paris. Connie Kay, whose work John Lewis had known with Lester Young, was brought in, with little preparation, for a job in Philadelphia. He has been with the group ever since.

The players have grown, even beyond expectation. Milt Jackson was obviously one of the great jazz vibists from his first appearance on the scene, but he has enlarged the range and subtlety of sounds he evokes, and his lovely choice of notes in improvising. He has also continued, obversely, to confirm and explore his own natural earthiness and swing.

John Lewis is one of the great accompanists in jazz, and that is particularly remarkable when one remembers that he prefers a spontaneous interplaying counterpoint to the more orthodox harmonic-percussive "comping" of most pianists. Lewis's solos are models of expressive economy and strongly communicated introspection and introversion—rare but not unknown qualities in this supposedly extravert music.

The present LP represents a return of the MJQ to Atlantic Records, the company with which it first affirmed and solidified its reputation, and for which it has recorded some of its best works.

Piazza Navona echoes back to the title piece of the Quartet's first Atlantic LP, *"Fontessa." Fontessa* was Lewis's first compositional tribute to the analogies between American jazz and Renaissance *commedia del l'arte.* "I had particularly in mind," he has said, "their plays which consisted of a very sketchy plot and in which the details, the lines, etc., were improvised."

Fontessa later became the full-length suite, *The Comedy* (also recorded for Atlantic), and the climactic section of *The Comedy* is *Piazza Navona,* named for a handsome, still-standing Roman square where *commedia* players might perform. This version is of course complemented by the dimension of an illustrious brass section (Snookie Young and Joe Newman, trumpets; Garnett Brown, trombone; Jim Buffington, French horn; Don Butterfield, tuba), for which Lewis has written so economically and effectively, letting it carry some of the burden of Renaissance allusion, while the Quartet extemporizes the modern American analogues. (Whatever happened to those strange, infrequent complaints that the Quartet wasn't earthy or funky or swinging or whatever?)

England's Carol also goes far back in the Quartet's and Lewis's repertory. Over the years, Lewis's version of this traditional melody has conveyed his love of its original European musical idiom, of counterpoint, and of jazz improvisation, in the several versions and settings in which he has performed it. This version has Lewis appropriately (and finally!) on harpsichord, and it is clearly the earthiest, most American reading yet. It has a new companion piece here in *Variations on a Christmas Theme* (another familiar one), again with brass and harpsichord, in a performance that is loose and free enough to make it seem to have been in the Quartet's repertory for a very long time.

Dancing somewhat resembles a couple of the Quartet's earlier pieces. But it is an interpretation of the contemporary boogaloo blues, with Lewis echoing and varying the kind of figures that the Ray Charles or James Brown brass sections shout out so superbly. *Dancing*'s eight-to-the-bar rhythm is, of course, as old as the most low-down, alley boogie woogie, and the Quartet plays as though it knows all about *that* too.

Listen also for Percy Heath during Lewis's solo. Clearly, here, on *Piazza Navona,* and on *Walkin' Stomp* [a jazz reggae, I should have added], Heath is taking on a markedly melodic, very effective, contrapuntal role in the ensemble.

Walkin' Stomp is something of an MJQ wonder, with Heath's bent-note riffing, and a complexity of texture and melody of the sort that usually has classical listeners exclaiming, "You mean they are *making this up?* As they *go along?*"

Trav'lin', on the other hand, is the sort of performance that is executed so well that it would surprise many listeners whether it were partly extemporized or not. A sort of relaxed *agitato* (I think you will forgive my apparent contradiction when you hear the piece), it has Lewis and Heath using the echo-plex.

Plastic Dreams should please jazz traditionalists (something the Quartet has been doing for a long time, by the way). It is a contemporary, varied approach to the idea of a jazz tango, an idea as old as Scott Joplin's ragtime tango *Solace,* the "Spanish tinge" of New Orleans jazz, and the second theme of *St. Louis Blues.*

I began by saying that the basics for the MJQ are four exceptional players who work superbly together. That is not quite enough, of course. One needs the guiding hand of a jazz composer who works collaboratively with the players. This LP is further evidence that the Quartet has that in John Lewis.

Lewis insists that he does not compose at the keyboard. If he tried, he explains, he would be carried away with the sheer pleasure of playing and never get to the composing. And that, it seems to me, might stand as a neat summary of how the Quartet as a whole feels about music. (*1971*)

Early Ornette

Ornette Coleman: The Shape of Jazz to Come*

(*These comments were written under the spell of a first discovery of Ornette Coleman's music. I have not tried to correct their occasional technical shortcomings here, although in passing I did add a couple of words which I felt clarified one point. They are offered here in part as a document of their times.*)

I believe that what Ornette Coleman is playing will affect the whole character of jazz music profoundly and pervasively. But I am not unique or original in believing that

* *"Ornette Coleman: The Shape of Jazz to Come,"* Atlantic 1317.

what he is doing is new and authentic. For two examples, Percy Heath, bassist with the Modern Jazz Quartet, has been praising him for over two years, often to deaf ears. "When I first heard Ornette and Don Cherry, I asked, 'What are they doing?,' but almost immediately it hit me. It was like hearing Charlie Parker the first time: it's exciting and different, and then you realize it's a really new approach and it makes a really valid music." And John Lewis, pianist and musical director of the Quartet, said last winter after hearing him, "Ornette Coleman is doing the only really new thing in jazz since the innovations in the mid-forties of Dizzy Gillespie, Charlie Parker, and those of Thelonious Monk."

What Ornette Coleman plays can be very beautiful, and it can have that rare quality of reaching out and touching each member of an audience individually. His melodies are unusual, but they have none of the harshness of self-conscious "experimental" borrowings. When we hear him he creates for our ears, hearts, and minds a new sensibility.

It is probably impossible for Ornette Coleman to discuss music without sooner or later using the word *love,* and he has said, with the innate modesty with which he seems to say everything, "Music is for our feelings. I think jazz should try to express more kinds of feeling than it has up to now." He knows well, then, the source and reason for his music. He also knows that he does not "own" it himself, nor "invent" it, but is responsible to something given to him. As is so necessary with an innovator in the beginning, he is not afraid to play whatever his Muse tells him to play: "I don't know how it's going to sound before I play it any more than anybody else does, so how can we talk about it *before* I play it."

What he has done is, like all valid innovations, basically simple, authentic, and inevitable—but we see that only once someone of a sublime stubbornness like Coleman's does it. The basis of it is this: if you put a conventional chord under my note, you limit the number of choices I have for my next note; if you do not, my melody may move freely in a far greater choice of directions. As he says of his improvising, "For me, if I am just going to use the changes themselves, I might as well write out what I am going to play." This does not mean that his music is "aharmonic" as is the music of a "country" blues singer, a Sonny Terry or a Big Bill Broonzy; nor that he invites disorder. He can work through and beyond the furthest intervals of the chords, and he has said, "From realizing that I can make mistakes, I have come to realize there is an order to what I do"—which, among other things, is as good a definition of maturity as I have ever heard.

As several developments in jazz in the last few years have shown, no one really needs to state all those chords that nearly everyone uses, and as some events have shown, if someone does state them or if a soloist implies them, he may end up with a harassed running up and down scales at those "advanced" intervals like a rat in a harmonic maze. Someone had to break through the walls that those harmonies have built and restore melody—but, again, we realize this only after an Ornette Coleman has begun to do it.

Since the world of jazz has been thickly populated with false prophets for the past ten years, Coleman may need his credentials, and it seems to me that among them are these: like the important innovators in jazz, he maintains an innate balance among rhythm and harmony and melodic line. In jazz, these three are really an identity, and any change in one of them without intrinsic reshuffling in the others invariably risks failure. Further, he works in terms of developing the specific, implicit resources of jazz, not by wholesale importations from concert music. Like most of the great ones, Coleman has a deep and personal feeling for the blues which is unmistakable. *Peace* seems to me a lovely example of that, and of how his playing adds to the emotional range of jazz. These things, and the fact that he breaks through the usual 32-, 16-, and 12-bar forms both in his compositions and in his improvising, all spring from an inner musical necessity, not from an outer academic contrivance. I think that, compositionally, *Congeniality* is an excellent introduction to many of these things, to the authenticity of his work and the way he is extending the whole idea of instrumental composition in jazz. Finally, to say (as some have) that the solos on such a piece do not have a relationship to his melodies is quite wrong. As a matter of fact, most jazz solos are not related to their theme-melodies, but to the chords with which the themes are harmonized. Coleman and Cherry may relate to the emotion, the pitch, the rhythm, the melody of a theme, without relating to "chords" or bar divisions. To a listener such relationships can have even more meaning than the usual harmonic ones.

Ornette Coleman started to play alto in Forth Worth, Texas, when he was fourteen. He remains largely self-taught on his instrument, but, inspired by a cousin, James Jordan, who was a music teacher in Austin, he studied books on harmony and theory quite early and thoroughly. One of his earliest influences was Red Connor, who had played with Charlie Parker and who, according to Coleman, played then the way Sonny Rollins and John Coltrane do now. Coleman has also heard and admired altoist Buster Smith, who now works in Dallas, and Smith was, as Parker said, "the guy I really dug."

Early jobs included carnival and rhythm and blues bands from which he was usually discharged for playing something the leader didn't like. Once in New Orleans a crowd smashed his instrument, apparently because *they* didn't like his playing. 1952 found him stranded in Los Angeles, where musicians at sessions would tell him he didn't know harmony and was out of tune. After a return to Fort Worth, he went back to Los Angeles in 1954; that was the crucial year for him, when he really broke through to the freedom that has characterized his work since. Still disapproved of, he supported himself and his wife Jayne, and soon a son, with day jobs while he stuck to his convictions about music.

In 1956, he met trumpeter Donald Cherry, who grasped what Coleman was doing, and began musically to breathe as he breathes. Cherry, who is six years younger than Coleman, moved from Oklahoma City to Los Angeles at four, and has gigged with professionals since 1951.

Bassist Charlie Haden is more than an asset to this recital; he too has grasped the essentials of Ornette's music and contributes to it. In his work here, I think one can see most clearly at least one of the things that Coleman's music has already achieved in group improvising. By abandoning conventional bass positions and the usual ideas of harmonic bass lines, he has found for himself and his instrument the area of pitch in which a piece and the improvisors are playing (Ornette sometimes wrote out a bass range for him). His bass forms a percussive-melodic part that participates much more directly in the music, rather than functioning as a contributing "accompaniment." Hear him especially on *Lonely Woman*. Finally, drummer Billy Higgins, who was brought into the group by Don Cherry, can shift tempos and play freely around the rhythm and time while maintaining the basic movement of, and swing of, a jazz beat.

The titles to these instrumental compositions are neither the usual ironies or throw-away, but are intended to mean what they say. *Congeniality,* for example, was originally named for a wandering preacher, but, Coleman says, "I'm not a preacher, I'm a musician, so I named it for what I think a *musician* feels toward an audience."

The tapes for this LP were edited during the third summer session of the School of Jazz at Music Inn in Lenox, Massachusetts, where Ornette and Donald Cherry were students. Ornette asked Gunther Schuller, who is a composer and French horn player in both jazz and classical idioms, practicing jazz critic, and instructor at the School, for help in the editing. Schuller's remarks on Ornette will follow, but because of what he says I will preface them by the note that his (Schuller's) own works have been praised for "one of the most immaculate senses of form in contemporary composition." He says, "Perhaps the most outstanding element in Ornette's musical conception is an utter and complete freedom. His musical inspiration operates in a world uncluttered by conventional bar lines, conventional chord changes, and conventional ways of blowing or fingering a saxophone. Such practical 'limitations' did not even have to be overcome in his music; they somehow never existed for him. Despite this—or more accurately, *because* of this—his playing has a deep inner logic. Not an obvious surface logic, it is based on subtleties of reaction, subtleties of timing and color that are, I think, quite new to jazz—at least they have never appeared in so pure and direct a form. Ornette's musical language is the product of a mature man who *must* speak through his horn. Every note seems to be born out of a need to communicate. I don't think Ornette could ever *play at* playing as so many jazzmen do. The music sits too deeply in him for that, and all these qualities are the more startling because they are not only imbued with a profound love and knowledge of jazz tradition, but are the first new realization of all that is implicit in the music of Charlie Parker." (*1959*)

Ornette Coleman: Free Jazz*

This is an exceptional record—exceptional in so many ways that it is hard to know where to begin. It is a continuous free improvisation with only a few, brief pre-set sections. It was done in one "take" at a single recording session. No one knew how long it would last; two tape machines were simply kept going, and when *Free Jazz* was over, it had taken over thirty-eight minutes—the length of an LP. There was nothing more to play, there were no re-takes, no splices.

Also, there is the unique instrumentation and personnel of the double quartet—two reeds, two trumpets, two basses, two drummers. There is the form of the piece. The fact that Ornette Coleman's style is also exceptional is well enough known to need no comment, but surely it has never been realized as it is here. Nor tested as it is here, for I don't suppose any jazz performance ever took bigger chances. Not only is the improvisation almost total, it is frequently collective, involving all eight men inventing at once. And there were no preconceptions as to themes, chord patterns, or chorus lengths. The guide for each soloist was a brief ensemble part which introduces him and which gave him an area of musical pitch. Otherwise he had only feelings and imagination—his own and those of his accompanists—to guide him. Ornette Coleman put it, "We were expressing our minds and emotions as much as could be captured by electronics."

Coleman has said that one of the basic ideas in his music is to encourage the improvisor to be freer, and not obey a pre-conceived chord-pattern according to set ideas of "proper" harmony and tonality: "Let's try to play the music and not the background." However, his point is basically

* *"Ornette Coleman: Free Jazz,"* Atlantic 1364.

emotional and aesthetic, not technical. The music should be directly and immediately "expressing our minds and our emotions rather than being a background for emotion."

Comparisons to contemporary non-objective painting, and certainly to contemporary Western music, surely spring to mind. So do comparisons to several analogous, time-honored practices in Eastern musics—the music of India, say—which are improvised according to the relative pitches and rhythms of the players, arrived at as they begin to play.

Often the "solo" here is (as in much New Orleans jazz) an exchange of the lead players. "The most important thing," says Coleman, "was for us to play together, all at the same time, without getting in each other's way, and also to have enough room for each player to ad lib alone—and to follow this idea for the duration of the album. When the soloist played something that suggested a musical idea or direction to me, I played that behind him in my style. He continued his own way in his solo, of course." A kind of polyphonic accompaniment based on pitch, melodic direction, an emotional complement, then.

There was little preparation for this session, except that the men had played together a couple of times. Coleman and Cherry are well used to such spontaneous, free improvising, of course, and drummers Billy Higgins, Ed Blackwell, and bassist Charlie Haden to playing with them. Scott LaFaro has played with the Coleman quartet too, but he has, he says, a deep and unshakable respect for harmony. Freddy Hubbard had played with Eric Dolphy, but Coleman heard him jamming with Don Cherry. You will hear that Hubbard's playing here sounds more conservative than the others. So does LaFaro's. I do not mean a negative criticism, and it is not that these men are any less willing to participate in total improvising. I think their ears hear harmonically almost automatically, and they may use not only the pitches of the brief ensemble introduction but immediately use their implied harmonies. It is a fascinating and effective contrast to some of the other soloists' sections.

Eric Dolphy's playing lies somewhere between. (Incidentally, he chose bass clarinet himself for this record from among the many reeds he plays.) He has said that he thinks of everything he does as tonal and harmonic, and he has a superb and sophisticated ear. Ordinarily he will use the harmonic framework of a piece and its choruses, but very freely, I think. Here he plays with still greater freedom, and with further development of his effort, "to get the instrument to more or less speak." (For example, something Hubbard plays at one point in his solo sets Dolphy's horn to sympathetic laughter behind him.)

Free Jazz is not a theme-and-variations piece in the usual sense. The written parts are brief introductions to each soloist, designed to introduce him and propel him musically. The soloists don't make "variations"; their improvisation is the music itself—the "theme" is whatever they invent, at the moment, in the act, of creation. The two bassists and the two drummers play throughout—brilliantly, I would say, and with spontaneous ears. Each horn man has a solo section of about five minutes. The leader's section is ten. The bassists each have five. The drummers share five. Some of the brief ensembles that introduce the soloists use approximately the same musical material, but sometimes it is fragmented, sometimes scored or voiced differently. Some ensembles use new material. Also, Coleman says that by a kind of spontaneous, nearly collective accident, one written introduction got left out during the playing and Blackwell got a repeat of Hubbard's ensemble part. The ensembles, by the way, do not necessarily come out "clean," with all the horns hitting the assigned notes exactly together. They are a joining of musical pitches which may seem almost casual at some points.

Free Jazz begins with brief polyphonic playing by the

horns—a kind of pitch and emotional "tune-up" for the piece, in effect. The first ensemble pitches follow. These are played in what Coleman calls a "harmonic unison"—an intuitive, homemade term (and procedure) which makes no sense until he explains that each horn has its own note to play but they are so spaced that the result will not sound like harmony but like unison. Dolphy's solo section follows. Notice how, at one point, when he doubles up his time, Dolphy gets an immediate response from the drummers. Coleman comments that Dolphy sometimes responds to his accompanists "as if he was playing all the instruments behind him."

The introduction to Hubbard's melodies follows. It is, Coleman explains, "in harmonic unison, but another degree." Hubbard begins with only the basses and drums, but soon he has encouraged comment from the other horns. Soon they have re-propelled him, in effect, and left him to the rhythm again.

The ensemble for Coleman's section begins with a gradual joining of sounds and a relatively more complex theme follows. Notice how Coleman re-phrases the theme twice before straying further into its implications. The group's responses to his melodies produce some of the most complex textures of the LP. His comment on hearing their interplay played back later was, "You can hear the others continue to build together so beautifully that the freedom even becomes impersonal."

Side 1 ends with Don Cherry's introduction, and fades out (electronically) as his solo begins. Side 2 begins with the ensemble again, a part of the "transposed unison" that also introduced Coleman. In the beginning the group comments, then Cherry is alone with the basses and drummers. Then fellow-trumpeter Hubbard returns. Then Coleman.

A partial repeat of the unison ensemble leads to Charlie

Haden's solo, as the drums swing quietly behind him. Haden is the melodist. LaFaro then follows a harmonic note by the group as the virtuoso bassist. When he has built to a complexity, LaFaro begins a contrasting, directly melodic, almost Django Reinhardt-like improvisation. (I am sure, by the way, that this recording would be exceptional—to use the word I began with—for the beautiful work of these two bassists alone.)

A "harmonic unison" note leads to Ed Blackwell. Coleman comments, "He plays a three-dimensional solo—rhythm and speed, then melodic rhythm, and the two together," in a fusion of essentials. Higgins follows another ensemble with his beautiful cymbal solo, with Blackwell behind, of course. Again, the technical and musical advances of this exceptional drumming have an emotional base; "the emotion he uses liberates the rhythms and metres," as Coleman puts it.

The ending is played "harmonically," with, as Coleman felt when he later heard the results, "the instruments sounding like voices singing."

Jazzmen have tried spontaneous group improvising without preconception before, of course—and almost invariably fallen into playing the blues within an acceptable key. It is surely a most telling tribute to the importance of this music that all of these young men, of different experience in jazz, were able to contribute spontaneously and sustain a performance like this one.

On the other hand, the man who isn't bothered about "newness" or "difference," but says only that "he sounds like someone crying, talking, laughing," is having the soundest sort of response to Ornette Coleman's music. (*1960*)

Ornette Coleman: Twins*

Just in case you're inclined to be ponderous or solemn about Ornette Coleman's music, go directly to *Joy of a Toy*. Hear the bursting humor of that well-titled melody. Hear the comic, cajoling, teasing fun in Ornette Coleman's solo. Hear those marvelous, convoluting phrases with which Don Cherry ends his improvising.

Ornette Coleman's music has always shown an expressive feel for the playful, the joyful, the whimsical side of human nature. And I'd say it's often there, deep down, in things he plays or writes in quite different moods.

But, no, that's not the way I should begin these notes. For there are some real treasures in this album. And I'd better get to telling you what they are—maybe with a little less subjectivity.

First Take is a real discovery, a long-forgotten performance, forgotten alike (it seems) by its producer, its performers, its engineer. It is, literally, a shorter "first take" of *Free Jazz,* one of the most important, provocative, and highly influential jazz recordings of the past fifteen years. Perhaps it was forgotten in the excitement over the way the full-length version turned out. But this shorter version has qualities and excitements of its own, and the validity therefore to stand on its own.

The general plan of *First Take* is the same as that of *Free Jazz*. A double quartet. Brief, loosely pre-set ensemble passages to introduce the soloists in turn. The solos have no pre-arranged structural limitations or barriers. But the soloists can be joined spontaneously by the other players as the spirit moves them, in a kind of improvised, atmospherically atonal counterpoint.

* *"Ornette Coleman: Twins,"* Atlantic SD 1588.

The loosely structured opening ensemble section beckons to Eric Dolphy on bass clarinet as the first soloist. His playing is looser here, more varied, sometimes perhaps even more "vocal" in quality than on *Free Jazz*. Notice too his shifts in tempo. And the fine way his conclusion calls in the second ensemble passage.

Freddie Hubbard, the next soloist, seems to me also more open and free here, and the soaring counterpoint provided by Hubbard and the others toward the end of his section is a wonder—the sort of thing one wants to hear again and again right after the first hearing.

The ensemble figures that usher in Ornette himself (like the following one that announces Don Cherry) make for a very Coleman-like melody. Coleman almost immediately sets up a medium-tempo, rocking swing against the faster clip the piece had taken before him, and that implied dual tempo continues throughout his remarkable section.

Don Cherry likes that medium, swinging mood too, for he re-establishes it in his section, in milder, quieter, less assertive terms.

In the section shared by the basses, the striking virtuosity of Scott LaFaro predominates over the simpler, more tradition-oriented melodies of Charlie Haden, and LaFaro continues to build after the brief, trumpet-led ensemble passage that subdivides the section.

I can't sort out the work of the drummers Billy Higgins and Ed Blackwell for you in these notes. But I'm sure that working on sorting them out would be fun after listening to the performance several times. (Just as it would be very revealing, also, to sort out which drummer is doing what with the original tempo and with the soloists' superimposed tempo during Coleman's section and Cherry's section.)

I don't think *First Take* quite reaches the darker depths or the shining heights of *Free Jazz,* but particularly for its sometimes looser and freer solos, and its terse, intense counterpoint, it is a very special and very valuable performance indeed.

I should not end these words without the sad reminder that two of its most valuable contributors, Eric Dolphy and Scott LaFaro, are dead.

Little Symphony, with its fine, unexpected start-and-stop theme, is from the session that contributed *Blues Connotation* and *Kaleidoscope* to the Atlantic LP called *"This Is Our Music"* and once again shows the difference in the music that was made by Ed Blackwell's fondness for a tight snare sound, as contrasted with Billy Higgins's on the earlier titles.

The title *Monk and the Nun* is, like all the titles here, a recently selected one. The Monk of the title is surely Thelonious—but the quality of the opening melody might tell you that without my help. The performance is still another from those three remarkable 1959 Los Angeles dates that contributed to *"The Shape of Jazz to Come"* and *"Change of the Century,"* that offered *Congeniality, Lonely Woman, Focus on Sanity, Ramblin'* and the rest, and that offered the LP titled *"The Art of the Improvisors."*

It seems that the further away we get from those first L.A. Atlantic Coleman dates, the larger they grow. I wish we had had *Monk and the Nun* on an LP years ago. Coleman's solo, with its sprightly opening accents, is wonderful—and wonderfully built, beginning to end. It is also another example of how the important jazzmen always reach back into the tradition and re-use and re-assess the most basic phrases and ideas in fresh and personal ways. And Don Cherry's musing, playful solo—well, it's no anti-climax, and that ain't easy under the circumstances.

Check Up comes from the date that produced the LP called *"Ornette,"* and like *R.P.D.D.* and the other pieces from that date, once had a title in initials before Coleman recently gave it this new one. *Check Up's* melody has a decidedly Southwestern, almost Mexican, quality (echoing Ornette's Texas upbringing, of course), and of all the Coleman originals I have ever heard, I think this one is the most directly reminiscent of other Coleman originals.

Notice how the bridge of the main melody is taken as a bass solo by LaFaro, but at *different* tempo. And notice too how, in the final statement of the main strain that closes the performance, Ornette leads the group into a tempo retard.

The leader's solo on *Check Up* is fittingly nostalgic, poignant, and a stretch-out. LaFaro's counterpoint behind him (as Blackwell quickly takes over the tempo-keeping chores) is a wonder. It is also an answer to those who think Scotty could not be effective when he played relatively simply, or think that his counter-melodies tended to crowd the soloist. The whole passage is a gem. And I like the way Cherry follows it by reaching into his down-home, country bag.

"Twins," by the way, is also Ornette's own title for the album. I haven't asked him why he chose it because I think I'd rather guess.

Maybe the twins are the members of the double quartet on *First Take* (in which case they are of course anything but identical twins). Maybe the twins are Coleman and Cherry, in the way that they state an opening theme: together, yet each making personal, loose interpretations that would not be allowed in the unison openings of the jazz of the 1940s and 1950s.

Or maybe the twins are the emotions and feelings and delights these men found in themselves and put into the music, and those you as a listener will get out of it. Hey—try that one and see. (*1971*)

V
WRITING
AND
READING

Criticism

A musician is supposed to have said recently that the criticism of jazz is a kind of joke and that there are no jazz critics. Without agreeing with him entirely, I am very sympathetic to his statement. But I say this to indicate that to me the words *critic* and *criticism* are rather special ones. A man who makes comments or reports on jazz records (or books, or plays, or movies) is not necessarily worthy of the title of critic.

The criticism of jazz is, like the criticism of any other art, "popular" or "fine," a kind of criticism. It is not a branch of publicity, nor a sideline of journalism. And a critical ability is not a natural consequence of an enthusiasm for jazz or of a knowledge of jazz, although it needs both of these things.

Philosophers would have us believe that criticism is a branch of philosophy, and some artists that it is a branch of creativity. But criticism has its own muse, and however much enlightment he constantly gets from both the philosopher and the artist, the critic needs a distinct, innate critical talent, a special sensibility and way of looking at things. His task is of an order much lower than that of either philosopher or artist, of course, but the ability he needs for his job is unique and uncommon, and a man either has it or doesn't have it. If the philosopher or artist (or journalist or historian) also has this critical ability, so much the better.

I think that the state of criticism of jazz in America is low, but I also think that the criticism of movies, plays, music in general, and painting is also low. Literature is lucky—it has a top level of criticism which is an excellent counter to the average American book review.

The innate critical ability is not enough in itself. It needs to be trained, explored, disciplined, and tested like any other talent.

If I recommended that this training should begin with Plato, Aristotle, and Lucretius and end with Eliot, Tovey, and Jung, I would not be saying something academic or pretentious but merely stating the most ordinary commonplace of Western civilization as it exists. And the critic should also know as much as he can of the best criticism being written around him in all fields.

But it is also the critic's business to be as perceptive and knowledgeable as he can. And critical perception (however much training it needs) is ultimately either there or not there.

The critic's questions are "How?" and "Why?" not merely "What?"

The points which follow come, with some changes, from Matthew Arnold. I present them not because I am especially interested in promoting Arnold's attitudes nor in promoting any "system," but because they seem to me to have something to say at this moment to the jazz writer and his reader.

1. The critic's first question is what is the work trying to do? Notice that this does not say, what do you think the artist *ought* to be trying to do. It also has little to do with a clairvoyant view inside the artist's head. And it assumes that the critic has observed the form of the piece and that he absorbed it with his feelings as well.

2. The second is, how well does it do it, and how and why so?

3. The third is, is it worth doing? Notice that this is the last question and *not* the first.

4. The critic should compare everything with the best that he knows whenever the comparison seems just and enlightening.

The questions are not easy, but no one ever said that criticism was easy, and even the very best critics can and will fail on at least some of them.

Ultimately, the critic makes a judgment, an evaluation. Value is based, in the final analysis, on feeling, not reason. But by feeling I mean a rational, conscious, individual function. I do not mean emotion which is irrational, impersonal, and can be irresponsible.

We have all heard it said that the criticism of jazz was once left to amateurs. That is not entirely true, nor is there any lack of amateurs today. But we do have now several writing about jazz who, although they really know what criticism is, don't know enough about music. On the other hand, there are some who know music, but don't know what criticism is. In jazz, of course, there is danger in knowing music since we are apt to apply the categories and standards of Western music rigidly and wrongly thereby. And there is also danger in knowing jazz: we may reject truly creative things because our knowledge of the past makes us think we know what a man *ought* to be doing—but that is true in any art.

The man who reviews jazz records has a terrible task: he can never, like his "classical" brother, judge an interpretation or performance against a norm because every jazz record is, in effect, a new work. Also, as George Orwell said of the hack book reviewer, day after day he must report on performances to which he has had little or no reaction worth committing to print—and that is true of the best critics and is neither a reflection on them nor necessarily on the music.

On the other hand, there could not possibly be as much true *creativity* in jazz as we are constantly told that there is, even though the medium is very much alive. How many novels, plays, poems, symphonies, paintings done in a year are *really* excellent?

And I wonder how many promising careers—and lives—

have been wrecked because of indiscriminate over-praise. I know of a few personally. Even if a musician is wise enough to discount what passes for criticism in jazz, he would have to be inhuman not to be somehow affected by it.

There is one job in jazz criticism that is neglected and which needs to be done, I think. It is also one which, since jazz is music and music the most abstract of the arts, is very difficult.

It is a better job on content and meaning. I am not opposed to technical analysis. We need more of that, too, and it can also help us with meaning, of course.

But especially now that jazz is so sophisticated, we need to talk frankly and honestly about what it is saying.

By an examination of content, I do not mean a kind of enthusiastic impressionism. Nor do I mean the kind of clever, chi-chi adjective-mongering we are all too familiar with. The critic's duty is accuracy and he should not sacrifice it for cleverness.

Of course, such an examination cannot be made with prejudice or pre-judgment. The first question is, what does this music express, not whether or not it *should* be expressing it.

The thing that separates listeners and commentators into "schools," I am convinced by the way, is not musical devices—passing chords, diminished ninths or sixteenth notes, or the lack of them—but the content that such devices enable a given style to handle. I think that jazz should be able to express as much as it can possibly learn to express in its own way.

Of course, the artistic and musical *expression* of emotion is not the same as its communication. A snarl, a sigh, a scream—these things communicate emotion, but they are not art, only a part of the raw material of art which the artist transforms.

I recommend this first, because greater consciousness is a part of growth in an art as well as in an individual.

I also recommend it because the appeal of jazz is still so very irrational, and I do not think it should be so much so any longer. (Of course, the appeal of all art is ultimately irrational, by definition, because it is art. But to many who like jazz, its appeal is almost entirely so.) It is the critic's business to make it less so, and unless he does, both he and jazz may be trapped. And dealing with content is the only way to give a good answer to that third question: is it worth doing?

As it is, we assure ourselves that jazz is an "art" and often proceed to talk about it as if it were a sporting event, an excuse for us to be verbally clever, a branch of big-time show biz, or an emotional outburst that affected us in a way we are not quite sure of. Perhaps we can at least do our best to create the kind of climate in which a jazz critic *could* function and which an art deserves. (*1958*)

Monk Goes to College

Anybody who gets into an institution of higher learning takes his chances. But works of art and the imagination take even bigger chances. Anyone who has ever heard a pedantic talk by a college professor on William Faulkner or read a piece of research scholarship on a play by Shakespeare knows about that.

So jazz is taking its chances these days in our universities and other institutions. But those of us who love the music and take it seriously know that it has to take them, because if any aspect of our culture needs and deserves the recognition of our future citizens and our future scholars, jazz does.

Still, I hope I may be allowed a shudder now and then at the results.

All of which is a kind of cautionary preamble to my saying that I recently read in a jazz textbook (never mind which one) that Thelonious Monk "seems not to have been influenced by anyone else, and his style of piano playing seems to have attracted no students."

Well now, you and I both know that there are passages in Ellington that are so like Monk, and vice versa, that novice listeners can't tell them apart.

And we know that Monk could not have played the way he did without also the example of Count Basie. (And the way Monk turned Basie's extravert humor completely inside out—well, yeah, there's that too.)

Then there are the very earliest recordings that have Monk—the Minton's things. They remind a lot of people of Teddy Wilson and maybe Clyde Hart.

As far as I'm concerned, not to be able to hear such things is like not being able to hear the affinity between Chopin and Liszt, Webern and Berg, or Rich Little and Richard Nixon.

But of course, if the listener *can't* hear any of this, the journalism and the biography of jazz will tell him. Everything I've said so far comes from the commonplaces of jazz writing of the last twenty-five years or so.

Now about Monk's effect on others: well, Bud Powell ended up playing quite differently, but again, if a listener can't *hear* the influence of Monk on him, Powell and his associates acknowledged it several times in interviews. And Powell has affected just about every pianist since 1945, one way or another.

At the same time, I have heard pianists like Randy Weston, Jaki Byard, Kenny Drew, Tommy Flanagan, and Barry

Harris use Monk very directly and on occasion even play a parody of Monk. Anyone for Mal Waldron? Andrew Hill? Cecil Taylor? (And speaking of Ellington, do you know that wonderful chromatic climb in the chord progression to *Rumpus in Richmond?* Then how about the bridge to Monk's *Well, You Needn't?*)

Of course, record reviews, interviews and other forms of journalism may not be substantial enough source material for a textbook writer—even one who can't hear how much Thelonious sounds like Duke. So maybe he should go to the critical essays on the man. They would be André Hodeir's and Gunther Schuller's (now, now, let's just leave my stuff out of this). There is no evidence that our textbook author has absorbed them, or even read them. Under the circumstances, I wouldn't be surprised if he didn't even know they existed.

But I'm not done yet. One of the most famous anecdotes of Monk's career has to do with the time he approved a "take" during a playback in a recording studio by remarking to Orrin Keepnews, "That's fine. That sounds like James P. Johnson."

But seriously, folks, I wonder just how responsible an academic is when he has not heard these things or read these things but he still presumes to comment—in effect, for posterity—on the talents and contributions of a man like Monk in an American textbook.

And just look at what you and I had to go through with to correct him! Personally I'm exhausted. And that's from just one little sentence—you should see the rest of this book!

As I say, when you go to college, you take your chances. (*1982*)

Biographies, Autobiographies, Profiles, and Oral History

The World of Duke Ellington, by Stanley Dance. Scribner's.

Combo: U.S.A.: Eight Lives in Jazz, by Rudi Blesh. Chilton.

Beneath the Underdog, by Charles Mingus (edited by Nel King). Knopf.

It has taken until recently (1969) for an important American musician (Gunther Schuller) to say in an academic setting (a conference at the University of California at Berkeley) that Edward Kennedy "Duke" Ellington is a major American composer.

He is, unequivocally, one of our greatest composers. But Ellington has been that perhaps since 1930, certainly since 1940. Yet most Americans, including most American musical academicians, if they really know of him at all, think of Ellington the popular song writer. They know of, let us say, *Sophisticated Lady* or *Mood Indigo* or *Don't Get Around Much Anymore* or *Do Nothing Till You Hear from Me.* They are less likely to know the instrumental compositions from which these sometimes simplified ditties were derived. And the great Ellington of, let us say, *Ko-Ko* or *Harlem Air Shaft* or *Blue Serge* or *Sepia Panorama* or *Bula* or *T.G.T.T.* remains largely unknown territory. Yet it is works like these that make him a great American composer.

The faculty leaders of our "stage" bands, ensembles that have become accepted presences on many of our high school and college campuses, introduce their charges and their audiences to our musical heritage with dilute Count Basie or, in more ambitious moods, with Stan Kenton. It is possible to admire Kenton, it seems to me, if (and perhaps only if) one has a knowledge of the conventions and devices of early twentieth-century European concert orchestration, an appetite for bombast and pretense, and little feel for art. Even a cursory study of the simplest Ellington (for example, *Blue Light*) or the ambitious Ellington (let us say, *Reminiscing in Tempo* or *A Tone Parallel to Harlem*) should put Kenton's musical posturing in proper aesthetic perspective.

The foregoing (part lamentation, part celebration) was set in motion by the arrival of one of the best examples of jazz journalism we have ever had, Stanley Dance's collection *The World of Duke Ellington*. Dance gives us Ellington and his music largely through a series of interviews and conversations. Four of them are semi-formal narratives by the maestro himself, twenty-nine are interviews with his musicians and associates, and three are accounts by Dance of such occasions as the preparation and performance of special works, a 1969 White House tribute to Ellington, and a Latin American tour.

On the whole, Dance has been thorough. He has (I believe) the only interview ever given by Ellington's great alto saxophonist, the late Johnny Hodges. He has spoken with such major Ellingtonians, past and present, as clarinetist Barney Bigard, trumpeter Charles "Cootie" Williams, trombonist Lawrence Brown, tenor saxophonist Ben Webster, trumpeter Clark Terry, as well as such recent and/or temporary Ellingtonians as bassist Jeff Castleman and singer Alice Babs.

There are gaps, to be sure, and some of them can't be filled. Trumpeter Bubber Miley's role in the orchestra was pivotal in the leader's early self-discovery of his own potential as a composer-orchestrator. But Miley was dead in 1932.

His counterpart on trombone, Joe Nanton, was dead in 1948. Also missing is the late cornetist Rex Stewart, who turned writer himself in his later years and left his own scripted heritage of Ellingtonia.

What emerges repeatedly, but always differently, in this volume is the ability of this exceptional man to discern the sometimes subtly hidden qualities and talents of each of his associates, to evoke and encourage those talents, and in so doing also to bring out the best in himself, and to produce thereby a body of music unique in human history.

There is constant direct contact with his subjects involved in Dance's volume, of course. There is less such involved in Rudi Blesh's *Combo: U.S.A.* which puts together a jazz combination of dubious musical compatibility.

Employing his own kind of earnest, enthusiastic, romanticized prose, Blesh has constructed biographical sketches of Louis Armstrong, soprano saxophonist Sidney Bechet, Jack Teagarden, tenor saxophonist Lester Young, singer Billie Holiday, drummer Gene Krupa, the innovative guitarist Charlie Christian, and pianist Eubie Blake, and he has largely used the familiar biographical sources and critical attitudes to do so.

True, there is a great deal of new material about Blake. Blake was a major ragtime composer in what one might call the Northeastern school (*Charleston Rag, Fizzwater,* etc.) and although he can be a charming performer and is a successful songwriter (*I'm Just Wild About Harry, Memories of You*), long ago chose that latter career and has made no direct contribution to jazz. There is some new material about Krupa too. Krupa is a famous man, but, granted that the point is debatable, one might name twenty drummers more musically important to jazz.

Combo U.S.A. might make a good biographical introduction to the people involved to anyone unfamiliar with the existing material. Otherwise, it seems to me an almost superfluous volume. And a quite disappointing one from the man whose contributions on the American arts have included a wonderful biography of Buster Keaton (called simply *Keaton*) and co-authorship of the invaluable *They All Played Ragtime.*

Blesh's earlier *Shining Trumpets,* however, was one of the most unfortunately influential volumes ever written on jazz. In it, the author, while giving pioneer praise to the work of early jazzmen from the teens and early twenties, at the same time heaped disapproval and even scorn on almost everything their successors had contributed to the music. Perhaps *Combo U.S.A.* is an effort to make up for that volume.

Incidentally, Blesh's inevitable source for his chapter on Sidney Bechet is Bechet's own *Treat It Gentle,* which, for at least half its length, is one of the most moving and enlightening autobiographies ever written by an American artist.

It would be impossible to say anything of fairness about Charles Mingus's *Beneath the Underdog* without first saying something about its author. Indeed one would like to say a great deal, but, in the interests of brevity, Mingus is one of the great bass virtuoso soloists ("the Segovia of his instrument," as one man has put it). He is also a man who, through his insight in assigning both a traditional percussive-harmonic accompaniment role to his instrument *and* a participating contrapuntal function to it as well, has changed every player's thinking about the bass in jazz. Let that much be said, aside from any discussion of his considerable talents as a composer.

Now. I am reviewing *Beneath the Underdog* frankly without having read all of it, and I did not read all of it because I did not think I could take any more of its implicit, grinding self-deprecation.

The surface of this book is sometimes musical—more so, I think, than other reviewers have admitted—but a great deal of it is sexual. And beneath the sometimes harsh, usually competitive, sex, beneath the armor of destructive (if thoroughly understandable) self-pity, there is a man who seems compelled to ask the world to hurt and misuse him, and who therefore shows the world his (and its) shadow side. And beneath *that* man is a being in torment who protests his accomplishments in music and his conflicting heritage as an American mulatto but who seems unable really to believe in them or to take confidence in them or take heart from them, and accept his place as an American musician.

The Gods have indeed looked with favor upon those complex and gifted men who, like Ellington, discover, accept, and live out their destinies from the beginning. (*1971*)

Grove American I: Not Just Missing Persons

My acquaintance with *The New Grove Dictionary of American Music* began when I looked up the vibraharp for some program notes I was working on. After an opening paragraph on the development of the instrument and its relationship to the marimba, the entry mentioned a 1933 piece by Milhaud and, later, another he had revised for vibraharp in 1947. There was mention of pieces by Boulez and Gunther Schuller (his *Paul Klee* sketches), and a late entry by Steve Reich was also there. In a short final paragraph, there was further comment on electronic amplification and a concluding sentence that credits jazz with having had several "virtuoso" performers on the instrument, naming Lionel Hampton, Milt Jackson, Red Norvo, and Gary Burton.

As Schuller and Reich know better than most of us, it is above all American jazz musicians who have developed the vibraharp, produced some of its masterpieces, and, it is reasonable to say, made Schuller's and Reich's work possible. It seems to me that a responsible entry in a work on American music should have acknowledged that, and perhaps told us how Hampton took a vocabulary of *arpeggios* and of standard drum figures and developed a style that could explore *Star Dust* as readily as *Flying Home.* Told us how Norvo transformed a highly developed xylophone into a vibrato-less vibes technique. How Milt Jackson explored an astonishing range of timbres, sonorities, vibrati, and dynamics on the most individual blues and the most sensuous ballads. (Are Jackson's crystalline *Moonrays* and *I Should Care* unworthy to stand beside Milhaud's *Marie's Announcement?* I think not!). How Walt Dickerson, with a carefully narrowed vibrato, and rubber mallets held close to their heads, discovered new timbres. And how Gary Burton can speedily manipulate as many as six mallets simultaneously (no vaudeville stunt this, all music). Indeed, by strictly conventional standards there are probably no virtuosi on the vibraharp besides Burton.

If ever a work called for an entry on the trumpet, surely it is a reference volume on American music. Our musicians, with two vibrato techniques, with lip sonorities and *glissandi,* half-value techniques, an array of mutes and combinations of mutes, a great extension of the working range (and not only on top), have radically changed the trumpet and its brass relatives. They have also made it a flexible carrier of melody that has left few composers or players unaffected world-wide. And in a reference work on American music, should I not also find some discussion of how our clarinetists and saxophonists have enlarged the range of their instruments? (Nowadays, Hamiett Bluiett plays comfortably in soprano range on baritone saxophone.)

After all, many books on European music will tell me what Paganini contributed to violin technique.

I also find some puzzling omissions and strange imbalances. Would I be carping to complain of no entries of Billy Page and Alex Hill? Possibly. Jo Garland and Buster Harding? I don't think so. Edgar Sampson, Frank Foster, and Bill Holman? Certainly not, yet none of these men is present.

Consider those who have written jazz history and criticism: among the many missing contributors are Albert Murray, Amiri Baraka (Leroi Jones), A. B. Spellman, Ralph Ellison, Stanley Crouch, and Bill Cole—and all of those I have named are black, by the way.

Consider some imbalances. Billy Strayhorn gets nineteen lines but Frank Zappa a page and a half. The entries on rock acts are quite *courant,* and some of them written in an enthusiastic prose that might fit in comfortably at *Tiger Beat* magazine: Blondie, the Talking Heads, Grace Jones, they are all here. But Don Pullen, David Murray, Arthur Blythe, Bobby McFerrin, and the World Saxophone Quartet are not to be found. Steve Allen is celebrated as a songwriter, but Fred Ahlert, Gus Arnheim, and Johnny Burke are not here. Pat Boone is here but Gene Austin is not.

In the January 13, 1987, *Village Voice,* there was a review of *The New Grove Dictionary of American Music* by Gary Giddins, which gives many more omissions and imbalances than I offer here, and I recommend that review to anyone seriously interested in American music. Giddins also does not disguise his disappointment and even bitterness, and I share those feelings. If the response is that, after all, a jazz *Grove* is in the works, well, assume that the job is done with excellence. That will still mean that jazz will be again put in a kind of cultural ghetto, will it not? And our understanding of all our music and its accomplishments can only be the poorer for that. (*1986*)

Grove American II: A Letter to a Friend

I have continued to ponder your comments on Grove American, and frankly I continue to be puzzled by them.

You say that the editors have chosen a broad scope for the book, and they have, but then you appear to be saying that critics, journalists, and other reviewers (including academics) may prove to be opposed to that decision. Gary Giddins in his review in the *Village Voice* is upset on the basis of length and proportion awarded to certain rock stars, and because of outstanding omissions in the jazz, blues, and popular entries. He is not opposed to the presence of other kinds of entries. He says, in effect, that the editors failed in the premise they set for themselves. I agree, and I add that they show themselves ignorant of some of the very ways Americans make music, and of the importance and significance of those ways throughout the world—perhaps an even greater failure.

You seem to feel that in taking a stand for all styles and genres of music the editors (and American musicologists in general) are taking a brave and innovative step. However, I know of very few real critics, and fewer good musicians, who don't know that music cannot be judged on the basis of genre or style. Ravel was *inspired* by Jimmy Noone! Stravinsky wrote that he composed his early pieces for clarinet under the spell of Sidney Bechet! And any comprehensive list of such influences and such prestigious respect for American musicians of their kind would be a long one.

I have heard you and others assert that in his book, *Yesterdays,* Charles Hamm questioned the long-held assumption that commercialism fatally compromised musical quality. I wonder who it was that held such an assumption, but we might examine it for a moment. Pushed just a bit further, it might mean that any just payment for artistic endeavor would be corrupting. Or it might imply a Marxist premise that competing in a corrupt capitalist society cripples artistic endeavor. Pushed in an opposite direction, it might imply that strange populist premise that pleasing a mass audience is the *only* path to artistic integrity. And frankly I have encountered all three assumptions in some comments on our music—sometimes side-by-side, with the resultant contradictions left unrecognized and unresolved.

In any case, have we forgotten that, by the 1920s, Gilbert Seldes, particularly, was telling us that he found some of our best and most original culture in our "lively arts," as he called them—in our popular musics, our movies, and our comic strips? Have we forgotten that, in a parallel case, our academics have by now accepted the movies as an art worthy of teaching and study?

More to the point, however, it seems to me that almost nowhere does Hamm show any interest in the musical quality of the songs he mentions. In *American Popular Song, 1900–1950,* on the other hand, Alec Wilder not only had a real interest, he had a bold commitment to musical quality. He seeks out the best songs in the idiom; he clarifies (in his cranky and sometimes awkward way) the issues and the musical standards, and he celebrates American song. He also makes me very proud of its accomplishments. Alec, God bless him, had critical and intellectual guts.

Is it also worth remembering that choosing to be broad and catholic in one's scope is not necessarily a virtuous position on the face of it? It might stem from sheer commercial opportunism (as it has with some of our record review publications). Or it might stem from a lack of discriminate taste. Or a basic cultural and aesthetic irresponsibility (shall I call it gutlessness?). Or it might stem from a real conviction and insight about our musics and their interrelationships and influence. But frankly I don't see much of that latter in Grove American.

As I said in my earlier comments to you, it seems to me that these volumes are not all "objective" but are full of implicit musical and critical judgments and unexamined assumptions. The dullest head among us could not fail to read the clear implication of such things as Frank Zappa's page and a half versus Billy Strayhorn's nineteen rather uninformative lines. And throughout, the real respect in Grove American repeatedly goes either to those who work within the European concert tradition (my example of the vibraharp should show that clearly enough) or to those who have huge popular audiences. No reader could miss those implications, it seems to me.

French scholars study all sorts of French concert composers, and every French conservatory student is required to know about almost all of them. But there is never any illusion about their relative musical merits and accomplishments. And the teachers and students understand that evaluating those composers and their works is a basically instructive part of the discourse.

If one is writing responsibly about any artistic endeavor, real objectivity is not possible, of course, and the need for clear aesthetic standards is inescapable, it seems to me. Any good literary historian, art historian, or theater historian knows that.

Have you never heard the theory, by the way, that the determination on the part of German academics somehow to proceed without evaluation or judgment paved the way for Hitler and for their egregious, tacit acceptance of the Nazis and their ways? It is an idea that has often been discussed.

All the best.

On Scholarship, Standards, and Aesthetics: In American Music We Are All on the Spot

(The essay which follows was originally read in briefer form before the Sonneck Society in American Music at its annual meeting in the spring of 1985.)

In a sense, I hesitate to undertake my subject with the Sonneck Society, which was originally organized on the premise that it approach American music informally, pleasurably, and however its membership seemed inclined.

Nevertheless, here we are with a learned quarterly, with annual meetings with several days of papers—in short, with all the paraphernalia of a society of scholars. And that in itself perhaps incurs certain obligations.

More important, we are a society devoted to a body of music, some of which has had considerable influence in the world, some of which has gained considerable respect, and some of which has an originality that might fill us with pride as Americans and with curiosity and dedication as scholars. Our music, in short, also implies certain responsibilities.

In this century the profession of musicology has strayed into a kind of relativism in which it is not so much concerned with musical values as with the "pure" research of historical fact and with valid conjecture. Whatever one feels about this turn of events—and about its influence in the classroom on the way music is presented to students—at least the hierarchy of European music, the membership in the pantheon of musical genius, is well enough established and relatively safe. And that is because major questions of evaluation have been settled and the major volumes of musical biography and music history have been written. If one picks up issues of the *Journal of the American Musicological Society* or *Musical Quarterly* or their European counterparts and finds no mention of Handel or Haydn or Brahms, one need not be afraid that they are forgotten or their reputations tarnished.

However, in American music there is no general agreement as to who or what has achieved musical excellence. Indeed, in American music scholarship we have no generally accepted standards of what constitutes excellence, what constitutes originality—even of what qualities might make American music American, or whether such qualities in themselves might be a basis for judgment.

One can also pick up issue after issue of *American Music* and see little or no indication that we have produced any major composers or that our culture can claim any musicians who, because of their uncompromising pursuit of musical excellence, should command our highest respect and most serious and detailed attention. One can attend panel after panel at the Sonneck Society's annual meetings and hear little discussion of major figures or of major events in any style, genre, or period of American music. And should there be a paper on a William Billings or a Virgil Thomson, the issues involved are all too often nonmusical or, in terms of any real evaluation of the man's life or works, peripheral or downright trivial.

So, it seems to me that an urgent intellectual need is not being met. But for us the need has the further urgency that we are still being told by some that our subject is unworthy of serious attention, that our music simply has not accomplished enough to make it a worthwhile subject for academic acceptance, research, and scholarship.

It seems self-evident that in this century, in particular, Americans have not only received and contributed to the mainstream of European music, we have also produced outstanding composers in idioms of our own. We have produced, for example, Charles Tomlinson Griffes and Elliott Carter, who have absorbed the practices of European schools—for Griffes, German Romanticism, then French Impressionism; for Carter, Viennese atonality—and used them to make music of commendably individual personality. And we have produced Charles Ives and Duke Ellington, each of whose music is so outstanding, so audaciously original, that it may be said to create standards of its own.

Now it is true that we acquire a formidable package from European music scholarship: a series of established masterpieces, established methods of research, standards of musical value, and even of allowable responses to music. And some of us are intimidated lest all this cut us off from the proper pursuit and appreciation of our own music in all its diversity.

But any real originality in art, any major artistic innovation, may force us to rethink our most fundamental assumptions on any art, even on occasion to rebuild the pantheon of artists itself, or at least add another wing. Besides, inside that "baggage" of European scholarship lie invaluable philosophical principles and guides by which any music, and any art, may be studied and evaluated.

Learning to deal with music, or with painting, poetry, or drama—with any aesthetic activity—is learning to deal with a subject that calls on us for high standards and exacting scrutiny, but for which there are no absolute standards and no unbreakable rules. Nevertheless, as the great music historians and critics have shown us, there is always a need to search for those standards that clarify artistic achievement. Mankind does not make music to satisfy the needs of social history.

As Wilfred Mellers wrote recently in a review of Charles Hamm's *Music in the New World,* "To put it crudely, what used to be called value is inescapable if one is to say anything worth the carrying about 'society' as a whole, for some music is 'better than' other music because it reveals more about the business of living. . . ." And, "genuine social history cannot be achieved without a measure of artistic understanding: which involves criticism and evaluation, however fallible."

I would like to suggest some specific problems and to offer some further convictions. And the first is clearly assumed in my quotations from Mellers.

1. I am puzzled and distressed by the growing use of populism, of yesterday's popularity, as a basis for evaluation. Could I, for example, do justice to Cole Porter through yesterday's "Top 40"? Cole Porter, who under the apparent simplicity of his songs knew so much about major-minor relationships and who worked them out on bass lines of such chromaticism that, of course, no untrained singer could possibly follow them, but which, of course, thousands of

untrained singers have followed correctly since they were written?

2. Are some of us perhaps unconsciously committed to a position that it really is better to be appreciative than discriminating? Do we assume that a person who knows too much or is too critical is a killjoy or a snob, and hopelessly cut off from the things most people like? Do some of us assume that in a true democracy creative excellence does not exist—or at least if it does, it would be elitist to recognize and celebrate it? I sometimes think that if literary scholars behaved the way some of us seem to be behaving, they would be discussing Edgar Guest with their poetry students and passing on Jacqueline Susann to future educated Americans.

3. I assume that the training of musical taste is a part of our responsibility as teachers and scholars. If I cannot give a student insight into how and why the *Trout* Quintet or, in my case, Ellington's *Ko-Ko* are great pieces of music, then I am not doing my job. And a student with an enthusiasm for a second-rate musician is the greatest challenge, and the greatest opportunity for real education, that a teacher could hope for.

4. If the European-derived standards and first principles of evaluation and judgment seem inappropriate or impossible to apply fairly to our music, then we need new ones. And we need to seek them out in the same way that the great music historians and critics of the past did: by discovering them in the accomplishments of the great musicians—in our case the great American musicians. However, if I were to discuss *Shenandoah* as a melody, how different would my standards be from those given me by Mozart and Schubert? Not at all different, I suspect.

5. I also have a strong conviction that anyone who deals with the arts at the graduate level should learn aesthetics and thereby learn how to deal effectively and fairly with

human endeavors, where standards are high but not absolute, where rules can be broken, and where the ultimate basis of value lies in human feeling, not human reason.

6. I have also never quite understood the classroom convention by which what is generally called "the background" of a work of art is thrust into the foreground and sometimes made the only ground. I was moved by Ives's *Concord* Sonata without having been told how its author earned his living. Only after I had been dazzled by Art Tatum's treatment of *Aunt Hagar's Blues,* and only after I had compared it with W. C. Handy's simple original, was I interested to learn that Tatum was blind and from Toledo. My first response to *Shenandoah* was that it was a magnificent melody. My second was to ask myself what I meant by that—what, after all, *is* good melody? And I was fascinated by the singular rhythmic development of Irving Berlin's *Putting on the Ritz* (going Gershwin's *Fascinatin' Rhythm* at least two better), and I was eager to analyze it, before I knew for whom he had written it and in what musical film of what year.

7. I think the questions of what our music has been like, who its major figures have been, and what its influence has been are questions we need to address. It is we, after all, who will be passing on a musical heritage to future generations—and, like it or not, evaluating that heritage is a basic and crucial aspect of our task. And it seems to me time that as a society of scholars we should encourage one another to address that task above all others. I suggest that at this stage of our endeavors, each of us should be addressing his or her energies to important figures and major events—critically, analytically, biographically, and in scholarly editions of scores.

8. The scholarly pursuit of any subject involves making several assumptions, the most basic of which is that one's subject is worth studying. Those who undertake scholarship

would do well therefore to ask themselves again and again how and why their subject deserves study and, as a corollary, how and why their particular approach is respectful, fruitful, and, in our case, musical.

In all of this, however, we should never forget that art is concerned with human experience and that music is human feeling rendered into sound and put into order. We should also remember Goethe's statement that we can understand nothing without love.

Not only Virgil Thomson, George Gershwin, and Duke Ellington but also John Philip Sousa, Carl Ruggles, and Harold Arlen did their best by their talents and by our country's need for its own musical culture. It is up to us to do our best by them.

If not that, then what are we here for? (*1985/86*)

*Why Aren't We Using the Classics?**

(*A somewhat longer version of this essay, written in collaboration with Doug Richards, appeared in* Jazz Educators Journal. *A different version, again written with Richards, appeared in the* Bulletin for Research in Music Education.)

In the late 1970s Big Three–United Artists Music had in its catalogue four classic Duke Ellington transcriptions, scores and parts, as entries in its "National Jazz Ensemble"

* Portions of this article are reprinted from the *Jazz Educators Journal* XIX, no. 2, the official publication of the National Association of Jazz Educators, Box 724, Manhattan, KS 66502.

series. And not too long thereafter, Big Three–United Artists decided it would publish no more Ellington.

That decision was surely a sad event in American music publishing. Here was an effort, an almost isolated effort, to offer authentic music by a man generally acknowledged to be not only the major jazz composer-orchestrator but, to many observers, a major twentieth-century composer regardless of musical category. And that effort had failed.

The Big Three decision becomes even sadder when we think about the reasons behind it. Nobody much was buying the orchestrations of *Ko-Ko, Harlem Air Shaft, Mainstem,* and the *Concerto for Cootie.* And if few people bought them, then few bands played them—not concert bands, not rehearsal bands, and (saddest of all) not student jazz ensembles.

Two related assumptions lie behind what makes the above so lamentable. The first is that student ensembles should be playing Ellington's works. The second assumption is that jazz education should be using the great works of the past to help train the music's future performers. The puzzling truth is that, for the most part, jazz education does not do that. One cannot go to an American student jazz concert and expect to hear at least *some* music from the Fletcher Henderson book, the Jimmie Lunceford Orchestra, among the many Ellington classics, or the early modern works of Dizzy Gillespie anywhere on the program. Indeed, if we do hear any Ellington, it is apt to be someone else's orchestration—and let's admit it, re-orchestrating a master like Duke makes as much sense as re-orchestrating Stravinsky.

The classical conservatories, of course, train their students with Bach, Mozart, Beethoven, and the other great works of the past. It might be argued that the conservatory is training musicians who will be playing Bach, Mozart, and

Beethoven. But those students are also going to play Corelli, Soleri, Glinka, and Offenbach, not to mention John Williams. Nevertheless, their core training comes from the great masters.

One reason that jazz education has proceeded the way it has probably comes from the experience of its teachers. In the real world of jazz every group, from big band to quartet, wants to make its own music—its own originals, its own versions of other's originals, its own interpretations on other people's chord changes or its own chord changes. But is that the whole story of the way in which jazz musicians have traditionally trained and in which they trained themselves?

Not really. Most players learned to solo first by imitating their idols—by ear, or by transcribing their solos from records, or both. The few more-or-less informal teachers who were around in the 1930s and 1940s taught that way. Lennie Tristano had Lee Konitz, Warne Marsh, and his other pupils spend hours with the records of Bix Beiderbecke, Lester Young, and others, and they did so willingly. Ambitious and promising young bands lifted the music from records for their charts, or played from big band stock arrangements (if—happily—they weren't too simplified from the originals).

In addition, recorded jazz was only about twenty-five years old when Charlie Parker first started to make records. Parker could not only get his Lester Young on records, he could hear him live. And he could hear Buster Smith, Coleman Hawkins, Don Byas, and Art Tatum this way and learn from them all. Among alto players, he knew the technical challenge of Jimmy Dorsey and the artistic challenge of Johnny Hodges. But Parker knew his Louis Armstrong too. (If you doubt that, listen to Parker's *Visa* solo on the "live" LP *"Bird at St. Nicks."* You'll hear him quote one of the Armstrong masterpieces note for note.)

Certainly our students should know how Parker himself made use of his influences. If they do, they will be even more aware of his quick harmonic imagination and ability to improvise a spontaneous, logical, structure that (on its own terms) is still unchallenged. But if students also know Parker's sources as well, what might they make of such a complex challenge? In any case, aren't we cheating them if we don't offer them the challenge and the knowledge?

There should be a few hundred Ellington works available to us all for teaching, for study, for performance; from the astonishing early twelve-man masterpieces like *Old Man Blues* and *Daybreak Express* at least through the selections on *"Afro-Bossa"* and *"Far East Suite."*

Certainly Ellington's music presents problems in interpretation (what good music doesn't?), but the challenge is not that the individual parts on most of the pieces are too difficult. And, in fact, the parts on the Henderson and early Basie masterpieces are simpler still. And wouldn't students be delighted to learn that they can play Gillespie's *Things to Come* when the 1946 band had difficulty with it when the style was new? Well, both the record and the chart are available.

No student ensemble has to train on trendy trash that will soon be forgotten. Remember, an educator is not there to confirm the taste which students already have. In jazz, as in any other subject, a teacher is there to enlarge and develop students' taste, and give them and their audiences a sound perspective on the past and its accomplishments.

If I seem to be suggesting that our student orchestras should be turned into something like cultural museums, well, we absolutely are not suggesting that. Bands should play new music, and play music written by their members and instructors. Individuality, originality, new challenges are the life blood of jazz, and an essential tool in teaching it.

Some of us, however, seem to have been acting as though jazz had no past worthy of serious attention. A music which has no past probably does not have much of a present. And it may have no future at all.

Classical players gain something from their studies beyond technical and artistic training. They learn to see themselves as a part of an illustrious musical continuity that goes back for centuries. They also learn to place themselves soundly in that continuity. By and large, jazz musicians have not had the sense of stability and worth that can come only from such a perspective. In recent years, jazz players have had only themselves, their contemporaries, and their immediate predecessors for their orientation. Is it not a major task of education to provide its future musicians with something larger and more stable?

To go back to where we started, get out your copies of *Ko-Ko, Mainstem, Harlem Air Shaft* and set your students to work on them. They'll be playing some of the best music ever written by an American. (*1987*)

Index